Tom Intrator
P24 Plasma Physics Group
Los Alamos National Laboratory
MS E526
Los Alamos, NM 87545
Email: intrator@lanl.gov

Principles of
Fusion Energy

Principles of Fusion Energy

An Introduction to
Fusion Energy for Students of
Science and Engineering

A. A. Harms
McMaster University

K. F. Schoepf
University of Innsbruck

G. H. Miley
University of Illinois

D. R. Kingdon
McMaster University

World Scientific
Singapore • New Jersey • London • Hong Kong

Published by

World Scientific Publishing Co. Pte. Ltd.

P O Box 128, Farrer Road, Singapore 912805

USA office: Suite 1B, 1060 Main Street, River Edge, NJ 07661

UK office: 57 Shelton Street, Covent Garden, London WC2H 9HE

British Library Cataloguing-in-Publication Data
A catalogue record for this book is available from the British Library.

ISBN 981-02-4335-9

Printed in Singapore.

Contents

Preface

Fusion energy is widely perceived as the ultimate terrestrial energy source. This appealing prospect has emerged by reason of scientific experiment, by the expectation of continuing scientific and technological progress, and by intriguing observations such as the following: life on earth is sustained as a consequence of fusion reactions in the sun; a medium-sized lake contains sufficient hydrogen fusion fuel to supply a nation with energy for centuries; the known diversity of fusion fuel cycles offers the eventual possibility of radiologically clean energy; fusion power appears essential for deep-space explorations; ... However, between this set of appealing notions and current scientific understanding and available technologies, there exists a barrier of considerable proportions. Both a broadly sustained community commitment and a high level of motivation by its participants is required for the realization of this ultimate source of energy.

As active research participants in several areas (both technical and geographic) of the fusion energy enterprise, we have long been sensitive to emerging technological perspectives related to this unique form of energy. In addition, as educators, we have repeatedly sought a conceptual and didactic framework for fusion lectures which would integrate the fundamentals of fusion phenomena with the successful experiences of the past and provide a link to the broader promise of emerging developments; additionally, and at a more subjective level, we have also sought an instructional balance between the pragmatic near-term educational role of a societal objective with the long-term inspirational value of a theme. Thus, while it is clearly essential to emphasize the well established concepts of magnetic and inertial confinement approaches to fusion, we believe it is also important to discuss topics such as spin-polarized fusion, advanced-fuel fusion reactions, muon-catalyzed fusion, and other related and emerging concepts. A synthesis which includes these and similar topics will, in our view, impart a most desirable perspective not only to the next generation of fusion scientists and nuclear engineers, but also to other professionals concerned with energy for the long term.

The writing of this text has been pursued, on and off, for nearly two decades. In retrospect, this has provided us with good time to accommodate the two divergent developmental paths which have become solidly established in the fusion energy community: the process of sequential tokamak development towards a prototype and the need for a more fundamental and integrative research approach before costly design choices are made. Our belief is that we have herein accommodated both interests in a coherent instructional format and the amount and level of material contained here allows for both avenues to be pursued.

In developing our subject we have found it useful to identify several distinct themes. The first is concerned with preliminary and introductory topics which relate to the basic and relevant physical processes associated with nuclear fusion. Then, we undertake an analysis of magnetically confined, inertially confined, and low-temperature fusion energy concepts. Subsequently, we introduce the important blanket domains surrounding the fusion core and discuss synergetic fusion-fission systems. Finally, we consider selected conceptual and technological subjects germane to the continuing development of fusion energy systems.

Our target group of interest is the senior undergraduate and beginning graduate university student in science or engineering. Familiarity with selected aspects of modern physics and a working knowledge of the differential calculus and vector algebra is assumed as a minimum prerequisite. In support of our pedagogical objectives, we have chosen to place considerable emphasis on the development of physically coherent and mathematically clear characterizations of the scientific and technological foundations of fusion energy specifically suitable for a first course on the subject. Of interest therefore are selected aspects of nuclear physics, electromagnetics, plasma physics, reaction dynamics, materials science, and engineering systems, all brought together to form an integrated perspective on nuclear fusion and its practical utilization. While the subject is of necessity broad, a focused pedagogical emphasis is consciously pursued: to identify and synthesize relevant physical concepts and their associated mathematical constructs and thereby provide a learning experience appropriate for subsequent more specialized work in any of the several areas of fusion energy.

In the course of our involvement in teaching and research in fusion energy, we sense a deep debt of appreciation to many with whom we have been in contact on matters of fusion energy. This includes numerous participants in various specific fusion energy programs at national and international research centres and colleagues at various universities. A particular word of thanks to Dr. D.P. Jackson (Chalk River Nuclear Laboratory, Canada), Prof. B. Lehnert (Royal Institute of Technology, Sweden), and Dr. G. Melese (General Atomics, USA) for their review of earlier drafts of this text. Additionally, we acknowledge those undergraduate and graduate students who, over the years, have passed on to us their comments on various versions of this work: A. Bennish, B. Bromley, B. Carroll, G. Cripps, B. Diacon, G. Gaboury, E. Hampton, X. Hani, T. Harms, S. Hassal, S. Ho, A. Hollen, M. Honey, J. Marczak, S. Mitchell, R. Ramon, P. Roberts, G. Sager, R. Scardovelli, A. Sguigna, P. Stroud, Y. Tan, D. Welch and J. Zielinski.

Our thanks to the several patient secretaries for typing the text and to the artists for the drawings. It is, however, with a particular sense of appreciation that we acknowledge two sterling co-ordinators for their accommodating disposition:

Jan Nurnberg (McMaster University) and Christine Stalker (University of Illinois).

To all we express our thanks.

A.A.H., K.F.S., G.H.M., D.R.K.
January 2000

PART I   CONTEXT, PHENOMENA, PROCESSES

1. Introduction

Matter and energy are fundamental components of our physical world. These components manifest themselves in a variety of ways under different physical conditions and can be affected by a variety of processes. Our interest in this first chapter relates specifically to the fusion of light nuclides which forms the basis of energy release in stars and which is expected to be harnessed on earth.

1.1 Matter and Energy

It is a common observation that matter and energy are closely related. For example, a mass of water flowing into the turbines of a hydro-electric plant leads to the generation of electricity; the rearrangement of hydrogen, oxygen, and carbon in chemical compounds in an internal combustion engine generates power to move a car; a neutron-induced splitting of a heavy nucleus produces heat to generate steam; two light nuclei may fuse and immediately break up with the reaction products possessing considerable kinetic energy. Each of these examples illustrates a transformation from one state of matter and energy to another in which an attendant release of energy has occurred.

These matter-energy transformations may be represented in various forms. For the hydro-electric process we may write

$$m(h_1) \rightarrow m(h_2) \tag{1.1}$$

where $m(h_1)$ and $m(h_2)$ is a mass of water at an initial elevation h_1 and final elevation h_2; the resultant energy E released can be evaluated by computing $(mgh_1 - mgh_2)$, where g is the local acceleration due to gravity.

An example of an exothermic chemical reaction is suggested by the process

$$CH_4 + 2O_2 \rightarrow 2H_2O + CO_2 \tag{1.2}$$

with an energy release of about 5 eV[*].

The case of neutron induced fission of a ^{235}U nucleus is represented by

$$n + {}^{235}U \rightarrow vn + \sum_i P_i \tag{1.3}$$

where n is a neutron, P_i is a particular reaction product, and n is the number of neutrons emitted in this particular process. Here, the total energy released possesses a slight dependence on the kinetic energy of the initiating neutron but

---

[*] Appendix A provides equivalents of various physical quantities.

is typically close to 200 MeV.

The fusion reaction likely to be harnessed first is given by

$$^2H + {}^3H \rightarrow n + {}^4He \qquad (1.4)$$

with an energy release of 17.6 MeV. Accounting for the fact that the above species react as nuclei, we assign in a more compact notation the names deuteron, triton, and alpha to the reactants and reaction product, to give

$$d + t \rightarrow n + \alpha. \qquad (1.5)$$

The fundamental features of matter and energy transformation are thus evident. In the hydroelectric case, a mass of water has to be raised to a higher level of potential energy–performed by nature's water cycle–and it subsequently attains a lower state with the difference in potential energy appearing as kinetic energy available to generate electricity. For the case of chemical combustion, an initial energetic state of the molecules corresponding to the ignition temperature of the fuel, has to be attained in order to induce a chemical reaction yielding thereupon new chemical compounds. The energy release thereby is due to the more tightly bound reaction product compounds with a slightly reduced total mass; such a mass defect is generally manifested in energy release–typically in the eV range for chemical reactions. In the case of fission, the initiating neutron needs to possess some finite kinetic energy in order to stimulate the rearrangement of nuclear structure; interestingly, the thermal motion of a neutron at room temperature is sufficient for the case involving nuclei such as ^{235}U. For fusion to occur, the reacting nuclei must possess sufficient kinetic energy to overcome the electrostatic repulsion associated with their positive charges before nuclear fusion can take place; the alternative of fusion reactions at low temperature is also possible and will be discussed later. Again, in the case of fission and fusion, the reaction products emerge as more tightly bound nuclei and hence the corresponding mass defect determines the quantity of nuclear energy release–typically in the MeV range.

Evidently then, a more complete statement of the above processes is therefore provided by writing an expression containing both matter and energy terms in the form

$$E_{in} + M_{in} \rightarrow E_{out} + M_{out} \qquad (1.6)$$

with the masses measured in energy units, i.e. multiplied by the square of the speed of light. The corresponding process is suggested graphically in Fig.1.1.

The depiction of Fig.1.1 suggests some useful generalizations. Evidently, a measure of the effectiveness and potential viability for energy generation by such transformations involves microscopic and macroscopic details of matter-energy states before and after the process. In addition, it is also necessary to include considerations of the relative supply of the fuel, M_{in}, the toxicity of reaction products, M_{out}, the magnitude of E_{out} relative to E_{in}, as well as other technological, economic and ecological considerations. Additional issues may include availability of the required technology, deployment schedules, energy conversion losses, management and handling of the fuel and of its reaction products, economic cost-benefit, environmental impact, and others.

Fig. 1.1: Schematic depiction of matter and energy flow in a matter-energy transformation device. The length of the arrows is to suggest a decrease in mass flow, $M_{out} < M_{in}$, with a corresponding increase in energy flow, $E_{out} < E_{in}$.

1.2   Matter and Energy Accounting

Conservation conditions are fundamental aids in the quantitative assessment of nuclear processes. Of paramount relevance here is the joint conservation of nucleon number and energy associated with an initial ensemble of interacting nuclear species of type a and type b which, upon a binary collisional interaction, yield two particles of type d and e:

$$a + b \rightarrow d + e . \qquad (1.7)$$

Note that the details of the highly transient intermediate processes are not listed– only the initial reactants and the final reaction products are shown.

An accounting of all participating nucleons is aided by the notation

$$_{Z}^{A}X = \begin{pmatrix} \text{nuclear species named X containing} \\ \text{A nucleons of which Z are protons} \end{pmatrix} \qquad (1.8)$$

and therefore yields the more complete statement for the reaction of Eq.(1.7) in a form which lists the number of nucleons involved in this nuclear rearrangement:

$$_{Z_a}^{A_a}X_a + _{Z_b}^{A_b}X_b \rightarrow _{Z_d}^{A_d}X_d + _{Z_e}^{A_e}X_e . \qquad (1.9)$$

Recall that A_j is the sum of Z_j protons and N_j neutrons in the nucleus, $A_j = Z_j + N_j$. Nucleon number conservation therefore requires

$$A_a + A_b = A_d + A_e \qquad (1.10)$$

and, similarly for charge conservation we write

$$Z_a + Z_b = Z_d + Z_e . \qquad (1.11)$$

Characterization of energy conservation for reaction (1.7) follows from the knowledge that the total energy E^* of an ensemble of particles is given by the sum of their kinetic energies E_k and their rest mass energies $E_r = mc^2$; here m is the rest mass of the particle and c is the speed of light in free space. The total energy of an assembly of particles, for which we add the asterisk notation, is therefore

$$E^* = \sum_j (E_{k,j} + E_{r,j}) = \sum_j (E_{k,j} + m_j c^2). \qquad (1.12)$$

For the nuclear reaction of Eq.(1.7) we write the total energy–which must be conserved–as

$$E^*_{before} = E^*_{after} \qquad (1.13a)$$

and hence

$$(E_{k,a} + m_a c^2) + (E_{k,b} + m_b c^2) = (E_{k,d} + m_d c^2) + (E_{k,e} + m_e c^2). \qquad (1.13b)$$

Rearrangement of these terms yields

$$(E_{k,d} + E_{k,e}) - (E_{k,a} + E_{k,b}) = [(m_a + m_b) - (m_d + m_e)] c^2. \qquad (1.14)$$

This important equation relates the difference in kinetic energies–before and after the collision–to their corresponding differences in rest masses; thus, as shown, a change in kinetic energy is related to a change in rest masses. Particle rest masses have been measured to a very high degree of accuracy allowing therefore the ready evaluation of the right-hand part of this equation. This defines the Q-value of the reaction

$$Q_{ab} = [(m_a + m_b) - (m_d + m_e)] c^2 = -[(m_d + m_e) - (m_a + m_b)] c^2 \qquad (1.15a)$$

and represents the quantity of energy associated with the mass difference before and after the reaction. Hence, we may write more compactly

$$Q_{ab} = (-\Delta m)_{ab} c^2 \qquad (1.15b)$$

where Δm is the mass decrement (i.e. $\Delta m = m_{after} - m_{before}$) for the reaction. Evidently, Q_{ab} is positive if $(\Delta m)_{ab} < 0$ and negative otherwise; the former case–involving a decrease of mass in the process–constitutes an exoergic reaction and the latter may be termed endoergic. Further, for the case of $Q_{ab} < 0$, the kinetic energy of the reaction-initiating particle must exceed this value before a reaction can be induced; that is, a threshold energy has to be overcome before the reaction will proceed.

Equation (1.15b) is a form of the famous relation

$$E = mc^2 \qquad (1.16)$$

and asserts–as first proposed by Einstein–that matter and energy are equivalent. As a consequence, we may assert that if processes occur which release energy of amount E, then a corresponding decrease in rest mass of amount $(-\Delta m)$ must have taken place.

1.3   Component Energies

A detailed kinematics characterization of reaction (1.7) requires the specification of both the kinetic energy and the momentum of the initial state of the reactants a and b as well as–depending upon the reaction details desired–the appropriate field forces which may act on the particles. However, some useful relations about the energies of the reaction products d and e can be obtained for the simple case in which the reacting particles possess negligible kinetic energies relative to the Q-value of the reaction, i.e. $E_{k,a} + E_{k,b} \ll Q_{ab}$, and in which the total energy liberated is shared by the two reaction products d and e in the form of their kinetic energy. Under these conditions, Eqs.(1.14) and (1.15a) give

$$\tfrac{1}{2}m_d v_d^2 + \tfrac{1}{2}m_e v_e^2 \approx Q_{ab} \; . \tag{1.17a}$$

Then, restricting this analysis to the case that the centre of mass be at rest, Fig.1.2, momentum conservation provides for

$$m_d v_d = m_e v_e \; . \tag{1.17b}$$

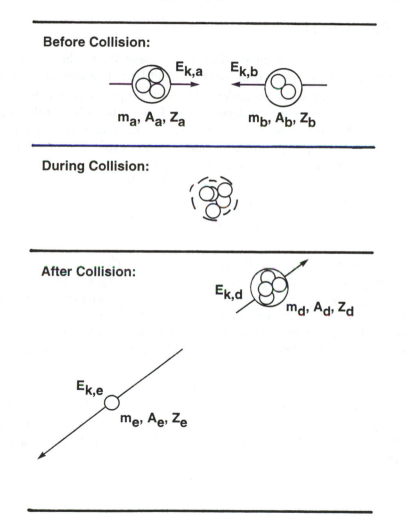

Fig. 1.2: Kinematic depiction of a head-on nuclear fusion reaction with the centre of mass at rest.

Solving Eq. (1.17b) for either v_d or v_e and substituting into Eq. (1.17a) yields kinetic energies $E_{k,d}$ and $E_{k,e}$ for the two reaction products:

$$E_{k,d} \approx \left(\frac{m_e}{m_d + m_e}\right)Q_{ab}, \qquad E_{k,e} \approx \left(\frac{m_d}{m_d + m_e}\right)Q_{ab}. \tag{1.18}$$

For the specific case of d-t fusion, Eq. (1.5), for which $Q_{dt} = 17.6$ MeV, the neutron and alpha particle kinetic energies are therefore found to be

$$E_{k,n} \approx \tfrac{4}{5}Q_{dt} \approx 14.1 \text{ MeV}, \qquad E_{k,\alpha} \approx \tfrac{1}{5}Q_{dt} \approx 3.5 \text{ MeV}. \tag{1.19}$$

Thus, an 80 - 20% energy partitioning occurs between the reaction products.

1.4   Fusion Fuels

Observations of natural and induced processes have shown that numerous types of fusion reactions for which Q > 0 can be identified. The variables for different reactions are the interacting nuclides[*], the reaction products which emerge, the Q-value of the reaction, and the dependence of the probability for the reaction to take place on the kinetic energy of the reactants. The fusion reaction most readily attainable under laboratory conditions and which is expected to be the first used for power generation purposes is the d-t reaction

$$d + t \rightarrow n + \alpha + 17.6 \text{ MeV}. \tag{1.20}$$

Another most accessible fusion reaction involves deuterium nuclei as fuel:

$$d + d \rightarrow \begin{cases} p + t + 4.1 \text{ MeV} \\ n + h + 3.2 \text{ MeV} \end{cases} \tag{1.21}$$

where h is chosen to represent the helium-3 nucleus ($^3\text{He}^{2+}$). This representation may appear somewhat unusual, but is seen to simplify notation in subsequent chapters. Equation (1.21) shows that d-d will fuse via two distinct reaction channels known to occur with almost equal probabilities at specific reaction conditions. Further, fuel deuterons may also fuse with two of the reaction products (tritons and helium-3) giving, in addition to the reactions of Eqs. (1.21) and (1.20),

$$d + h \rightarrow p + \alpha + 18.3 \text{ MeV}. \tag{1.22}$$

The above fusion reactions involve deuterons and the successively more massive light nuclides. Continuing along this pattern, a large number of reaction channels have been identified in specific cases of which d-^{6}Li fusion is an example:

---

[*]Appendix B displays the light-nuclide part of the Chart of the Nuclides.

$$d + {}^6Li \rightarrow \begin{cases} {}^7Be + n + 3.4 \text{ MeV} \\ {}^7Li + p + 5.0 \text{ MeV} \\ p + \alpha + t + 2.6 \text{ MeV} \\ 2\alpha + 22.3 \text{ MeV} \\ h + \alpha + n + 1.8 \text{ MeV} \end{cases} \quad (1.23)$$

Here, each reaction channel possesses a unique probability of occurrence.

Fusion reactions involving the lightest nucleus, that is the proton, may occur according to the processes

$$p + {}^6Li \rightarrow h + \alpha + 4.0 \text{ MeV} \quad (1.24a)$$

$$p + {}^9Be \rightarrow \begin{cases} \alpha + {}^6Li + 2.1 \text{ MeV} \\ d + 2\alpha + 0.6 \text{ MeV} \end{cases} \quad (1.24b)$$

$$p + {}^{11}B \rightarrow 3\alpha + 8.7 \text{ MeV} \quad (1.24c)$$

as well as others. Some reactions based on t and h are

$$t + t \rightarrow 2n + \alpha + 11.3 \text{ MeV} \quad (1.25a)$$
$$h + h \rightarrow 2p + \alpha + 12.9 \text{ MeV} \quad (1.25b)$$

and

$$t + h \rightarrow n + p + \alpha + 12.1 \text{ MeV} . \quad (1.25c)$$

Several features associated with fusion reactions need to be noted. First, the physical demonstration of a fusion reaction is not the only consideration determining its choice as a fuel in a fusion reactor. Other considerations include the difficulty of bringing about such reactions, the availability of fusion fuels, and the requirements for attaining a sufficient reaction rate density.

Another feature of the various fusion reactions listed above needs to be emphasized: in each case a different fraction of the reaction Q-value resides in the kinetic energy of the reaction products. Thus, a fusion reactor concept based on high-efficiency direct energy conversion of charged particles would appear particularly suitable for those reactions which are characterized by a high fraction of the Q-value residing in the kinetic energy of the charged particles. This is of particular interest because the neutrons appearing as fusion reaction products invariably induce radioactivity in the materials surrounding the fusion core.

Third, the fusion fuels are evidently the light nuclides displayed on the Chart of the Nuclides. In a subsequent chapter we will show that a subatomic short-lived particle called a muon and produced in special accelerators, may also play a role as a fusion reaction catalyst.

Most current fusion research and development activity is based on the expectation that the d-t reaction, Eq. (1.20), will be used for the first generation fusion reactors. While the world's oceans as well as fresh water lakes and rivers contain an ample supply of deuterium with a particle density ratio of $d/(p+d) \sim 1/6700$, tritium is scarce; it is a radioactive beta emitter with a half life of 12.3

years, with the total steady state atmospheric and oceanic quantity of tritium produced by cosmic radiation estimated to be on the order of 50 kg. Since a 1000 MW_t plant will burn about 250 g of tritium each operating day, a station inventory in excess of 10 kg will be required for every d-t based central-station fusion power plant so that other sources of tritium fuel are required.

The main source of tritium is expected to be its breeding by capture of the fusion neutron in lithium contained in a blanket surrounding the fusion core. The relevant reactions in ^6Li and ^7Li are

$$n + {}^6Li \rightarrow t + \alpha \qquad (1.26a)$$

and

$$n + {}^7Li \rightarrow t + \alpha + n \qquad (1.26b)$$

with the latter possessing a high energy threshold $E_{thresh} \approx 2.47$ MeV. Lithium-6 and lithium-7 are naturally occurring stable isotopes existing with 7.5% and 92.5% abundance, respectively, and exist terrestrially in considerable quantity.

Additional sources of tritium may involve its extraction from the coolant and moderator of existing fission reactors, particularly heavy water reactors, where tritium is incidentally produced by neutron capture in deuterium via

$$n + {}^2H \rightarrow {}^3H . \qquad (1.27)$$

Of course, tritium could also be produced by placing lithium into control and shim rods of fission reactors.

Reaction (1.26b) is particularly interesting because the inelastically scattered neutron appearing at lower energy can continue to breed more tritium. Thus, in principle, it could be possible in such a system to produce more than 1 triton per neutron born in the d-t reaction. Indeed, present concepts for d-t reactors generally assume a lithium-based blanket surrounding the fusion core that allows for tritium self-sufficiency. These and additional concepts will be discussed in subsequent chapters.

1.5   Fusion in Nature

While a very small number of fusion reactions occur naturally under existing terrestrial conditions, the most spectacular steady state fusion processes occur in stellar media. Indeed, the formation of elements and the associated nuclear energy releases are conceived of as occurring in the burning of hydrogen during the gravitational collapse of a stellar proton gas; the initiating fusion process is

$$p + p \rightarrow d + \beta^+ + \nu + 1.2 \text{ MeV} \qquad (1.28)$$

where β^+ represents a positron and ν a neutrino. Then, the deuteron thus formed may react with a background proton according to

$$p + d \rightarrow h + 5.5 \text{ MeV} . \qquad (1.29)$$

Subsequently, this helium-3 reaction product could fuse with another helium-3 nucleus to yield an alpha particle and two protons:

$$h + h \rightarrow \alpha + 2p + 12.9 \text{ MeV} . \tag{1.30}$$

The next heavier element is beryllium, produced by

$$h + \alpha \rightarrow {}^{7}Be + 1.6 \text{ MeV} \tag{1.31}$$

and is an example of a rare helium-4 fusion reaction. Also, lithium may appear by

$$ {}^{7}Be + \beta^{-} \rightarrow {}^{7}Li + 0.06 \text{ MeV} . \tag{1.32}$$

A progression towards increasingly heavier nuclides is thus evident. This process is known as nucleosynthesis and provides a characterization for the initial stages of formation of all known nuclides.

Closed fusion cycles have also been identified of which the Carbon cycle is particularly important:

$$ {}^{12}C + p \rightarrow {}^{13}N + 1.9 \text{ MeV} $$

$$ {}^{13}N \rightarrow {}^{13}C + \beta^{+} + v + 1.5 \text{ MeV} $$

$$ {}^{13}C + p \rightarrow {}^{14}N + 7.6 \text{ MeV} $$

$$ {}^{14}N + p \rightarrow {}^{15}O + 7.3 \text{ MeV} \tag{1.33}$$

$$ {}^{15}O \rightarrow {}^{15}N + \beta^{+} + v + 1.8 \text{ MeV} $$

$$ {}^{15}N + p \rightarrow {}^{12}C + \alpha + 5.0 \text{ MeV} . $$

This sequence of linked reactions is graphically depicted in Fig.1.3 and may be collectively represented by

$$4p \rightarrow \alpha + 2 \beta^{+} + 2v + 25.1 \text{ MeV} \tag{1.34}$$

if all the reactions of Eq. (1.33) proceed at identical rates. This relation suggests that protonium burns due to the catalytic action of the isotopes ^{12}C, ^{13}C, ^{13}N, ^{14}N, ^{15}N, and ^{15}O.

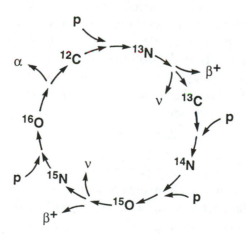

Fig. 1.3: Graphical depiction of the Carbon fusion cycle.

Problems

1.1   Determine the energy released per mass-of-atoms initially involved for a chemical process, Eq.(1.2), for nuclear fission, Eq.(1.3), for nuclear fusion, Eq.(1.5) and for a water molecule falling through a 100 m elevation difference in a hydroelectric plant.

1.2   Calculate the reaction Q-values for each of the two branches of the d-d fusion reaction, Eq.(1.21).

1.3   What fraction of the original mass in d-t fusion is actually converted into energy? Compare this to the case of nuclear fission, Eq.(1.3).

1.4   Calculate the kinetic energies of the reaction products h and a resulting from p-^{6}Li fusion, Eq.(1.24a), ignoring any initial motion of the reactants.

1.5   Calculate the total fusion energy, in Joules, residing in a litre of water if all the deuterons were to fuse according to Eq.(1.21).

1.6   Redo problem 1.5 including the burning of the bred tritium according to Eq. (1.20).

1.7   Consult an astronomy text in order to estimate the mean fusion power density (Wm^{-3}) in the sun; compare this to a typical power density in a fission reactor.

1.8   The first artificial nuclear transmutation without the use of radioactive substances was successfully carried out in the Rutherford Laboratory by Cockcroft and Walton when they bombarded Lithium (at rest) with 100 keV proton canal rays (protons accelerated by a voltage of 100 kV and passing through a hole in the cathode). By scintillations in a Zincblende-screen, the appearance of α-particles with a kinetic energy of 8.6 keV was determined.
    (a) Formulate the law of energy conservation valid for the above experiment referring to the nuclear reaction

$$_3^7Li + {}_1^1H \rightarrow 2\ _2^4He$$

and find therefrom the reaction energy, Q_{p7Li}, via the involved rest masses (m_p, m_α= 6.64455 × 10^{-27} kg, m_{7Li} = 11.64743 × 10^{-27} kg).
    (b) In the Cockcroft-Walton experiment, conservation of momentum was proven by cloud chamber imaging whereby it was observed that the tracks of the two α-particles diverge at an angle of 175°. What angle follows from the law of momentum conservation by calculation?
    (c) A further reaction induced by the protons in natural lithium is

$$_3^6 Li + {}_1^1 H \rightarrow {}_2^4 He + {}_2^3 He$$

Provide an argument that shows the α-particles detected in the Cockcroft-Walton experiment cannot stem from this reaction.

2.   Physical Characterizations

A number of fusion reactions of interest were listed in the preceding chapter but little reference was made to the conditions under which these reactions might occur. We now consider the fusion process itself and some characterizations of conditions which are fundamental to an understanding of controlled nuclear fusion.

2.1   Particles and Forces

The general fusion reaction, Eq.(1.7) and Fig.1.2, may be more completely characterized by noting that an unstable intermediate state may be identified in nuclear reactions. That is, we should write

$$a + b \rightarrow (ab) \rightarrow d + e + Q_{ab} \qquad (2.1)$$

where (ab) identifies a complex short-lived dynamic state which disintegrates into products d and e. The energetics are determined according to nucleon kinetics analysis, with nuclear excitation and subsequent gamma ray emission known to play a comparatively small role in fusion processes at the energies of interest envisaged for fusion reactors.

Two-body interactions can be examined from various perspectives. For example, Newton's familiar law of gravitational attraction applies to any pair of masses m_a and m_b to yield a force

$$\boldsymbol{F}_{g,a} = -G \, \frac{m_a \, m_b}{r^3} \boldsymbol{r} \qquad (2.2)$$

effective on particle a. Here, G is the universal gravitational constant and $\boldsymbol{r} = \boldsymbol{r}_a - \boldsymbol{r}_b$ is the displacement vector between the two interacting particles, while r denotes its absolute value. While this force expression is universal, a simple calculation will show that for nuclear masses of common interest, this force is significantly weaker than the electrostatic and nuclear forces associated with nuclides and hence can be neglected.

The important electrostatic force between two isolated particles of charge q_a and q_b separated by a distance r in free space is determined by Coulomb's law, given by

$$\boldsymbol{F}_{c,a} = \frac{1}{4\pi \, \varepsilon_o} \frac{q_a \, q_b}{r^3} \boldsymbol{r} \qquad (2.3)$$

for the electrostatic force felt by particle a; here ε_o is the permittivity of free space and the factor 4π is extracted from the proportionality constant by reason of convention. This force–repulsive for like charges and attractive for unlike charges–is of considerable importance in fusion.

From the definition of work and the phenomenon of energy stored in a conservative field, the work done in moving a particle of charge q_a from a sufficiently distant point to within a distance r of a stationary charge of magnitude q_b, is the potential energy associated with the resultant charge configuration. Specifically, this is given by

$$U(r) = \int_{\infty}^{r} \boldsymbol{F}_{c,a}(\boldsymbol{r}')d\boldsymbol{r}'$$

$$= \int_{\infty}^{r} \frac{1}{4\pi\,\varepsilon_o} \frac{q_a q_b}{(r')^3} \boldsymbol{r}' \cdot d\boldsymbol{r}' \qquad (2.4)$$

$$= \int_{r}^{\infty} \frac{1}{4\pi\,\varepsilon_o} \frac{q_a q_b}{(r')^3} (-r' dr')$$

$$= \frac{1}{4\pi\,\varepsilon_o} \frac{q_a q_b}{r}$$

subject to the restriction that the particle distance of separation r satisfies $r \geq R_a + R_b$ where R_a and R_b are the equivalent radii of the two charged particles. For nuclides of like charge, the potential energy at approximately the distance of "contact" $R_o = R_a + R_b$, is called the Coulomb barrier and, in view of Eq.(2.4), is given by

$$U(R_o) = \frac{1}{4\pi\,\varepsilon_o} \frac{q_a q_b}{(R_a + R_b)} . \qquad (2.5)$$

On the basis of electrostatic force considerations only, this then is the minimum kinetic energy an incident particle would have to possess in order to overcome electrostatic repulsion and come close enough to another particle for the short-range nuclear forces of attraction to dominate. For deuterium ions, this energy can be calculated to be about 0.4 MeV, depending upon the precise value for R_o. A useful approximation is $R_o \approx R_p(A_a^{1/3} + A_b^{1/3})$ where $R_p = (1.3\text{-}1.7) \times 10^{-15}$ m denotes the radius of a proton which cannot be assigned a definite edge for quantum mechanical reasons.

Consideration of quantum mechanical tunneling provides for a non-vanishing probability of penetrating the Coulomb barrier with energies less than $U(R_o)$. The probability for this penetration varies as

$$Pr(\text{tunneling}) \propto \frac{1}{v_r} exp\left(-\gamma \frac{q_a q_b}{v_r}\right) \qquad (2.6)$$

where v_r is the relative speed of the moving particles and γ is a constant. Thus, even at very low energy, a nucleus possesses a small, though finite, probability of compound formation with another nucleus. This compound can decay into fusion products and hence, some fusion reactions will also occur at room temperature, though at an insignificant rate.

At sufficiently small distances, $r < R_o$, the attractive strong nuclear force dominates and a compound nuclear state is formed. The kinetic energy of the initiating particles together with the resultant nuclear potential energy is then shared by all the nucleons. Nuclear stability considerations thereupon determine if and how the nucleus disintegrates. Figure 2.1 provides a graphical representation of these effects.

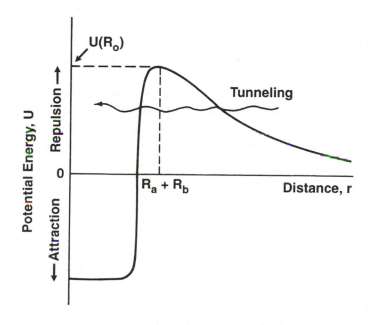

Fig. 2.1: Depiction of the ion-ion electrostatic repulsive potential for $r > R_o$ and nuclear attraction for $r < R_o$.

2.2   Thermal Kinetics

The recognition that the Coulomb barrier for light ion fusion is in the 0.4 MeV range for the lightest known nuclides–p, d, and t–suggests that one approach to the attainment of frequent fusion reactions would be to heat a hydrogen gas up to a temperature for which a sufficient number of nuclei possess energies of relative motion in excess of $U(R_o)$, Eq.(2.5). However, because of the tunneling effect, such excessive heating is not required and substantial rates of fusion reactions are achieved even when the average kinetic energy of relative motion of the reactant nuclei is in the tens-of-keV range. Note, however, that even providing such reduced conditions in a gas will be associated with heating up to temperatures near ~ 10^8 K, for which, in recognition of the ionization potential of hydrogen being only 13.6 eV, the gas will completely ionize. The result therefore is an electrostatically neutral medium of freely moving electrons and positive ions called a plasma. With the average ion energy in a fusion reactor plasma thus allowed to be substantially less than the Coulomb barrier, the energy released from fusion reactions must exceed–for a viable energy system–the total energy initially supplied to heat and ionize the gas, and to confine the plasma thus produced. In practice, this means that when a sufficiently high plasma temperature has been attained, we have to sustain this temperature and confine the ions long enough until the total fusion energy released exceeds the total energy supplied. In subsequent chapters, we will consider some details of the relevant phenomena and processes and identify parametric descriptions of energy balances.

On the basis of the above considerations then, it is apparent that selected aspects of the classical Kinetic Theory of Gases, augmented by electromagnetic force effects, can be used as a basis for the study of a plasma in which fusion reactions occur. Thus, for the case of N atoms of proton number Z_i, complete ionization yields N_i ions and, in the case of charge neutrality, $Z_i N_i = N_e$ electrons; that is

$$N \rightarrow N_i + N_e = N_i + Z_i N_i \; . \tag{2.7}$$

For hydrogenic atoms, $Z_i = 1$ and hence $N_e = N_i$.

Often, the expression "Fourth State of Matter" is also assigned to such an assembly of globally neutral matter containing a sufficient number of charged particles so that the physical properties of the medium are substantially affected by electromagnetic interactions. Indeed, such a plasma may also exhibit collective behaviour somewhat like a viscous fluid and also possesses electrostatic characteristics of specified spatial dimension.

For a state of thermodynamic equilibrium, the Kinetic Theory of Gases asserts that the local pressure associated with the thermal motion of ions and electrons is given by

$$P_i = \tfrac{1}{3} N_i m_i \overline{v_i^2} \tag{2.8a}$$

and

$$P_e = \tfrac{1}{3} N_e m_e \overline{v_e^2} \tag{2.8b}$$

where the subscripts i and e refer to the ions and electrons respectively, and $N_{()}$ refers to the subscript-indicated particle population densities; note that it is the average of the squared velocity which appears as the important factor. The average kinetic energy of the electrons and ions can be introduced by simple algebraic manipulation of the above equations:

$$P_i = \tfrac{2}{3} N_i \left[\tfrac{1}{2} m_i \overline{v_i^2} \right] = \tfrac{2}{3} N_i \overline{E_i} \tag{2.9a}$$

and

$$P_e = \tfrac{2}{3} N_e \left[\tfrac{1}{2} m_e \overline{v_e^2} \right] = \tfrac{2}{3} N_e \overline{E_e} . \tag{2.9a}$$

Similar expressions may be written for the neutral particles in a plasma.

Accepting the kinetic theory of gases as a sufficiently accurate description allows for the use of well-known distribution functions. Implicit in this assumption is that the plasma under consideration is sufficiently close to thermodynamic equilibrium and that processes such as inelastic collisions, boundary effects and energy dependent removal of particles are of secondary importance.

The distribution functions for the particles of interest include dependencies on space, time and either velocity, speed or kinetic energy. Here we take a stationary ensemble of N^* particles uniformly distributed in space–either neutrals, ions or electrons–allowing us to write

$$N(\xi) = N^* M(\xi) \tag{2.10}$$

where ξ is one of the independent characteristic variables of motion–**v**, v, or E– and $M(\xi)$ describes how the particles are distributed over the domain of this variable. Hence, $N(\xi)$ is the distribution function of the ensemble of relevant particles in ξ-space and the symbol $M()$ represents a normalized distribution function for the variable ξ such that

$$\int_{-\infty}^{\infty} M(\xi)d\xi = 1 \tag{2.11a}$$

with the integration performed over the entire definition range of the variable considered, i.e.

$$\int_{-\infty}^{\infty} M(\mathbf{v})d\mathbf{v} = \int_{0}^{\infty} M(v)dv = \int_{0}^{\infty} M(E)dE = 1 . \tag{2.11b}$$

Note that the particle number of the ensemble is given by

$$\int_{-\infty}^{\infty} N(\xi)d\xi = N^* \int_{-\infty}^{\infty} M(\xi)d\xi = N^* . \tag{2.12}$$

For a gas or, for the case of interest here, a plasma in thermodynamic equilibrium and in the absence of any field force effect, its particles of mass m moving in a sufficiently large volume follow the Maxwell-Boltzmann velocity distribution function given by

$$M(\mathbf{v}) = \left(\frac{m}{2\pi kT} \right)^{3/2} \exp\left(\frac{-\tfrac{1}{2}mv^2}{kT} \right) . \tag{2.13}$$

Here k is the Boltzmann constant and T is the absolute temperature for the ensemble. For the case of isotropy, the Maxwellian distribution of speeds is a generally satisfactory characterization and is given by

$$M(v) = \left(\frac{2}{\pi}\right)^{1/2} \left(\frac{m}{kT}\right)^{3/2} v^2 \exp\left(\frac{-\frac{1}{2}mv^2}{kT}\right), \quad 0 < v < \infty. \tag{2.14}$$

Finally, the corresponding distribution of particles in kinetic energy space E is described by the Maxwell-Boltzmann distribution

$$M(E) = \frac{2}{\sqrt{\pi}} \left(\frac{1}{kT}\right)^{3/2} E^{1/2} \exp\left(-\frac{E}{kT}\right), \quad 0 < E < \infty. \tag{2.15}$$

These three functions are illustrated in Fig.2.2.

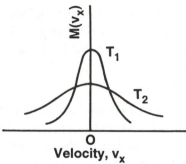

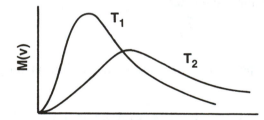

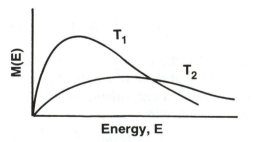

Fig. 2.2: Schematic depiction of the Maxwell-Boltzmann distribution functions for v_x, v and E as independent variables; in each case, $T_1 < T_2$.

2.3 Distribution Parameters

It is most important to recognize that while the particles possess a range of velocities, speeds, and energies, the temperature T describes a particular distribution function and is a fixed parameter for a given thermal state; changing the temperature of the medium will alter the various moments of the function but its characteristic shape is retained, Fig.2.2.

Further, in a volume domain containing a mixture of particles–as in the case of a plasma containing electrons, various ion species, and neutrals–each particle species may possess a different distribution function characterized by a different temperature. Then, however, the entire plasma is not in thermodynamic equilibrium. Indeed, in the presence of a magnetic field, even the same species may have a different temperature in, say, the direction parallel to the magnetic field lines than in the perpendicular direction. Several methods or devices used to obtain fusion energy involve plasmas that are just that–not in thermodynamic equilibrium. Most that will be considered herein, however, are not so and thus we will rely on Maxwell-Boltzmann distributions to characterize many of the plasmas that will be discussed in subsequent chapters.

Having a sufficiently accurate distribution function is of considerable utility. For example, the most probable value $\hat{\xi}$ –that is the peak of the distribution–is found by differentiating and finding the root of

$$\frac{\partial M(\xi)}{\partial \xi}\bigg|_{\xi=\hat{\xi}} = 0 . \tag{2.16}$$

For the three distribution functions listed, Eqs. (2.13), (2.14) and (2.15), this yields the following:

$$\frac{\partial M(\mathbf{v})}{\partial v_x}\bigg|_{v_x=\hat{v}_x} = 0 , \quad \hat{v}_x = 0 \tag{2.17a}$$

$$\frac{dM(v)}{dv}\bigg|_{v=\hat{v}} = 0 , \quad \hat{v} = \left(\frac{2kT}{m}\right)^{1/2} \tag{2.17b}$$

and

$$\frac{dM(E)}{dE}\bigg|_{E=\hat{E}} = 0 , \quad \hat{E} = \tfrac{1}{2}kT . \tag{2.17c}$$

In Eq. (2.17a), the subscript x is to suggest any one component of the vector $\mathbf{v}$; hence $\hat{\mathbf{v}} = \mathbf{0}$.

Average values can similarly be found based upon the formal definition of

$$\bar{\xi} = \frac{\int \xi M(\xi)d\xi}{\int M(\xi)d\xi} . \tag{2.18}$$

Thus, for the three cases of interest here we get

$$\bar{v}_x = \int_{-\infty}^{\infty} v_x \, M(\mathbf{v}) \, dv_x = 0 \quad (i.e. \ \bar{\mathbf{v}} = \mathbf{0}) \tag{2.19a}$$

$$\bar{v} = \int_{0}^{\infty} v M(\mathbf{v}) \, dv = \left(\frac{8kT}{m\pi}\right)^{1/2} \tag{2.19b}$$

and

$$\bar{E} = \int_{0}^{\infty} E M(E) \, dE = \tfrac{3}{2} kT \tag{2.19c}$$

with the particles possessing three degrees of freedom.

The analysis leading to the depictions of Fig. 2.2 makes it clear that the temperature T–here in units of degrees Kelvin, K–is an essential characterization of a Maxwellian distribution; hence, the numerical value of T uniquely specifies an equilibrium distribution. It has also become common practice to multiply T by the Boltzmann constant k and to call this product the kinetic temperature, which is obviously expressed in units of energy, either Joules (J) or electron volts (eV) with the latter generally preferred. Using this product kT, a Maxwellian population at T = 11,609 K may be said to possess a kinetic temperature of 1 eV; similarly, a 3 keV plasma in thermodynamic equilibrium has an absolute temperature of 3.48×10^7 K.

The convention of interchangeably using energy and temperature, wherein the adjective "kinetic" and Boltzmann's constant in kT are commonly suppressed, may seem peculiar, but expressing a physical variable in related units is a very common practice. For example, travelers often use time as a measure of distance (s = vt) if the speed of transport is understood, test pilots often speak of a force of so many g's (F = mg), and physicists often quote rest masses in units of energy ($E = mc^2$).

This convention of using the product kT leads to a number of uses which need to be distinguished; we note here several common cases:

kT = (kinetic) temperature of a plasma;

$\tfrac{3}{2} kT$ = average energy of Maxwellian-distributed particles;

$\tfrac{1}{2} kT$ = most frequently occurring particle energy of Maxwellian-

distributed particles;

$\sqrt{\dfrac{8}{m\pi}} \sqrt{kT}$ = average particle speed of Maxwellian-distributed particles.

2.4   Power and Reaction Rates

The power in a fusion reactor core is evidently governed by the fusion reaction rate. If only one type of fusion process occurs and if this process occurs at the rate density R_{fu} with Q_{fu} units of energy released per reaction then the fusion power generated in a unit volume is given by

$$P_{fu} = R_{fu} Q_{fu} \ . \qquad (2.20)$$

With R_{fu} expressed in units of reactions$\cdot m^{-3} \cdot s^{-1}$ and Q_{fu} in MeV per reaction, the units of the power density P_{fu} are MeV$\cdot m^{-3} \cdot s^{-1}$ which can be converted to the more commonly used unit of Watt (W) by the conversion relationship $1 \ eV \cdot s^{-1} = 1.6 \times 10^{-19}$ W since $1 \ W = 1 \ J \cdot s^{-1}$. For the case of a uniform power distribution the total energy released during a time interval τ in some volume V follows from Eq.(2.20) as

$$E_{fu}^{*} = V \int_{0}^{\tau} P_{fu} \, dt \qquad (2.21a)$$

and, at any time t

$$P_{fu} = \left(\frac{dE_{fu}^{*}}{dt} \right) \frac{1}{V} \ . \qquad (2.21b)$$

That is, energy may be viewed as the area under the power curve while power may be interpreted as an instantaneous energy current.

The energy generated in a given fusion reaction, Q_{fu} in Eq.(2.20), is the "Q-value" of the reaction, Eq.(1.15), and can be experimentally determined or extracted from existing tables; this part of the power expression is simple. However, the determination of the functional form of the reaction rate density R_{fu} is more difficult but must be specified if the fusion power P_{fu} is to be computed.

In order to determine an expression for the fusion reaction rate density, consider first the special case of two intersecting beams of monoenergetic particles of type a and type b possessing number densities N_a and N_b, respectively, Fig. 2.3.

In a unit volume where the two beams intersect, the number of fusion events in a unit volume between the two types of particles, at a given time, is given by a proportionality relationship of the form

$$R_{fu} \propto N_a N_b v_r \qquad (2.22)$$

where v_r is the relative speed of the two sets of particles at the point of interest. This relation is based on a heuristic plausibility argument for binary interactions. Obviously, some idealizations are contained in Fig.2.3 and Eq.(2.22), such as particles of varying energy and direction of motion as well as the interaction of particles with others of their species not being accounted for, but will be considered in subsequent sections.

The proportionality relationship of Eq.(2.22) can be converted into an explicit equation by the introduction of a proportionality factor represented here by σ_{ab} for a given v_r:

$$R_{fu} = \sigma_{ab} (v_r) N_a N_b v_r \ . \qquad (2.23)$$

The subscript ab and the functional dependence on v_r, indicated in $\sigma_{ab}(v_r)$, is to emphasize that the magnitude of this parameter is specifically associated with the particular types of interacting particles and their relative speed. The common name for $\sigma_{ab}(v_r)$ is "cross section" and, since all the terms of Eq.(2.23) are already dimensionally specified, its units are those of an area. This parameter has been assigned the name "barn", abbreviated b, and defined as

$$1 \, b = 10^{-24} \, cm^2 = 10^{-28} \, m^2 \, . \tag{2.24}$$

Figure 2.4 illustrates the cross section for a case of deuterium-tritium fusion. Note a maximum of a few barns in the $v_r = 3 \times 10^6 \, ms^{-1}$ range in this figure.

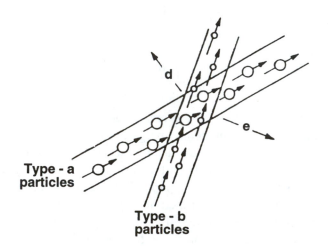

Fig. 2.3: Intersection of two particle beams resulting in fusion reactions a + b → d + e.

2.5   Sigma-V Parameter

The fusion reaction rate density expression of Eq.(2.23) is very restrictive since all particles were taken to possess a constant speed and their motion was assumed to be monodirectional. However, the general case of an ensemble of particles possessing a range of speeds and moving in various directions can be introduced by extending Eq.(2.23) to include a summation over all particle energies and all directions of motion. The integral calculus is ideally suited for this purpose requiring, however, that we redefine some terms. Letting therefore the particle densities be a function of velocity **v** endows them with a range of energies and range of directions; that is, we progress from simple particle densities which give the number of particles per unit volume, to distribution functions describing how many particles in a considered position interval move with a certain velocity, according to

$$N_a \to N_a(\mathbf{v}_a) = N_a F_a(\mathbf{v}_a) \qquad (2.25a)$$

and

$$N_b \to N_b(\mathbf{v}_b) = N_b F_b(\mathbf{v}_b) . \qquad (2.25b)$$

The terms which have replaced the previous particle densities are distributions in the so-called position-velocity phase space. Thus, the velocity distribution functions $F_a(\mathbf{v}_a)$ and $F_b(\mathbf{v}_b)$ satisfy the normalization

$$\int_{\mathbf{v}_a} F_a(\mathbf{v}_a) d^3 v_a = 1 , \qquad (2.26a)$$

as well as

$$\int_{\mathbf{v}_b} F_b(\mathbf{v}_b) d^3 v_b = 1 . \qquad (2.26b)$$

Here $d^3 v_{()}$ is to indicate integration over the three velocity components. Hence, though we show here only one integral, the implication is that for calculational purposes there will be as many integrals as there are scalar components for each of the vectors $\mathbf{v}_a$ and $\mathbf{v}_b$.

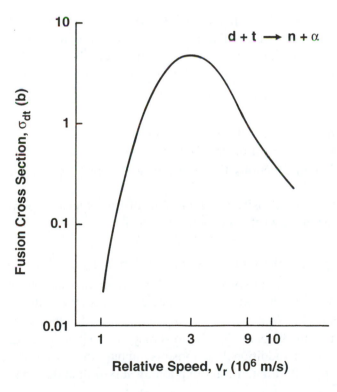

Fig. 2.4: Fusion cross section for a deuterium beam incident on a tritium target.

The relative speed of two interacting particles at a point of interest is, by its usual definition, given as

$$v_r = |\mathbf{v}_a - \mathbf{v}_b| .$$ (2.27)

Finally then, a general rate density expression for binary reactions follows by extension of Eq.(2.23) over the corresponding velocity space:

$$R_{fu} = \int\limits_{\mathbf{v}_a \mathbf{v}_b} \int \sigma_{fu}\left(|\mathbf{v}_a - \mathbf{v}_b|\right)|\mathbf{v}_a - \mathbf{v}_b|\, N_a\, F_a(\mathbf{v}_a)\, N_b\, F_b(\mathbf{v}_b)\, d^3 v_a\, d^3 v_b$$

$$= N_a\, N_b \int\limits_{\mathbf{v}_a \mathbf{v}_b} \int \sigma_{fu}\left(|\mathbf{v}_a - \mathbf{v}_b|\right)|\mathbf{v}_a - \mathbf{v}_b|\, F_a(\mathbf{v}_a)\, F_b(\mathbf{v}_b)\, d^3 v_a\, d^3 v_b .$$ (2.28)

By first impressions, this double integral appears formidable, particularly when it is realized that the vectors $\mathbf{v}_a$ and $\mathbf{v}_b$ possess, in general, three components so that a total of six integrations are required. However, aside from any algebraic or numerical problems of evaluation, this integral contains two important physical considerations. First, the cross section $\sigma_{fu}(|\mathbf{v}_a - \mathbf{v}_b|)$ must be known as a function of the relative speed of the two types of particles, and second, the distribution functions $F_i(\mathbf{v}_i)$ must be known for both populations of particles.

We note however that Eq.(2.28) possesses all the properties of an averaging process in several dimensions; that is, it represents averaging the product $\sigma_{ab}(|\mathbf{v}_a - \mathbf{v}_b|)|\mathbf{v}_a - \mathbf{v}_b|$ with two normalized weighting functions $F_a(\mathbf{v}_a)$ and $F_b(\mathbf{v}_b)$ over all velocity components of $\mathbf{v}_a$ and $\mathbf{v}_b$. Such averaging yields the definition

$$< \sigma v >_{ab} = \int\limits_{\mathbf{v}_a \mathbf{v}_b} \int \sigma_{ab}\left(|\mathbf{v}_a - \mathbf{v}_b|\right)|\mathbf{v}_a - \mathbf{v}_b|\, F_a(\mathbf{v}_a)\, F_b(\mathbf{v}_b)\, d^3 v_a\, d^3 v_b .$$ (2.29)

This parameter, here named sigma-v (pronounced "sigma-vee"), is often also called the reaction rate parameter. Note the implicit dependence on temperature via the distribution functions $F_a(\mathbf{v}_a)$ and $F_b(\mathbf{v}_b)$ so that $<\sigma v>_{ab}$ is a function of temperature.

The reaction rate density involving two distinct types of particles a and b, Eq.(2.28), is therefore written in compact form as

$$R_{fu} = N_a\, N_b < \sigma v >_{ab} .$$ (2.30)

This sigma-v parameter for the case of d-t fusion under conditions in which both the deuterium and tritium ions possess a Maxwellian distribution, that is $F_{()}(\ ) \rightarrow M_{()}(\ )$ in Eq.(2.29), and where both species possess the same temperature, is depicted in Fig.2.5.[*] Note that generally $<\sigma v>_{ab}$ will also be a function of space and time because the velocity distributions may also depend upon these variables.

We make two additional comments about the reaction rate density expression, Eq.(2.30). First, the assumption of Maxwellian distributions, i.e.

---

[*] Appendix C provides a tabulation of this and other $<\sigma v>_{ab}$ parameters.

$F_{()}(\ ) \to M_{()}(\ )$ as discussed in Sec.2.3, is very frequently made in tabulations of sigma-v; the reason for this is because many approaches to the attainment of fusion energy rely upon the achievement of plasma conditions that are close to thermodynamic equilibrium. For cases where equilibrium conditions do not exist, the appropriate distribution functions $F_a(\mathbf{v}_a)$ and $F_b(\mathbf{v}_b)$ must be determined and used in Eq.(2.29). For example, in some experiments, deuterium beams are injected into a tritiated target or into a magnetically confined tritium plasma to cause fusion by "beam-target" interactions. In such cases, one substitutes $F_d(\mathbf{v}_d)$ by a delta distribution function at the velocity $\mathbf{v}_b(t)$ characteristic of the instantaneous velocity of the beam ions slowing down in the plasma. However, $F_t(\mathbf{v}_t)$ for the plasma target could well be assumed to be Maxwellian at the temperature of the tritium plasma. Then the averaged product of σ and v_r is often called the beam-target reactivity $<\sigma v>_{dt}^{b}$ for d-t fusion and is displayed in Fig.2.6; it appears to be a function of both the target temperature and the instantaneous energy of the slowing-down beam deuterons.

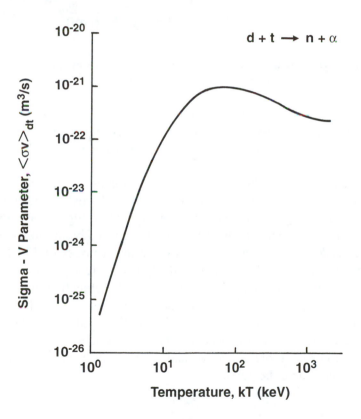

Fig. 2.5: Sigma-v parameter for d-t fusion in a Maxwellian-distributed deuterium and tritium plasma.

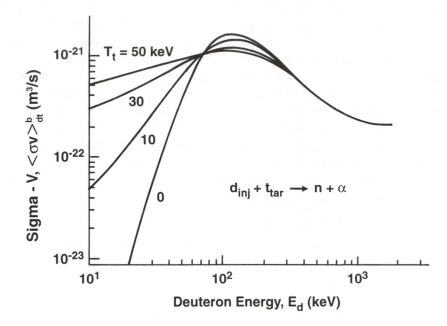

Fig. 2.6: Beam target sigma-v parameter for the case of deuterium injection into a tritium target of various temperatures.

Secondly, Eq.(2.30) assumes that the a-type and b-type particles are different nuclear species; the case when they might be indistinguishable is treated in a subsequent chapter where, in addition to d-t fusion, the case of d-d fusion is also considered.

A demonstration of the occurrence of fusion reactions is readily accomplished in a simple experiment employing a small accelerator which bombards a tritiated target with deuterium ions of such a high energy that during slowing down in the target they pass through the most favourable energy range for fusion; consider, in particular, the tritium plasma target of Fig.2.6 for this purpose. The appearance of neutrons and alphas as reaction products at the proper energies is then the proof of fusion events. The objective of fusion energy research and development is, however, the attainment of a sufficiently high fusion reaction rate density under controlled conditions subject to the overriding requirement that the power produced be delivered under generally acceptable terms. This is a considerable challenge and to this end a variety of approaches have been and continue to be pursued.

Problems

2.1   Calculate the ratio of gravitational to electrical forces between a deuteron and a triton.

2.2   Determine the Coulomb barrier for the nuclear reactions d-t, d-h, and p-^{11}B.

2.3   Confirm the correctness of Eqs.(2.17) and (2.19).

2.4   For Maxwellian distributed tritons at 9 keV, calculate
   (a) the average kinetic energy,
   (b) the average speed, and
   (c) the kinetic energy derived from the average speed of (b). Compare the energies of (a) and (c), and explain any difference.

2.5   Transform M(v), Eq.(2.14), into M(E), Eq.(2.15), with the aid of the appropriate Jacobian.

2.6   Find M(E) from M($\mathbf{v}$) for the case of isotropy using spherical co-ordinates.

2.7   Consult appropriate sources to determine particle densities N (m^{-3}), the corresponding energies kT (eV) and temperatures T (K) for the following plasmas: outer space, a flame, the ionosphere, commonly attainable laboratory discharges and values expected in a magnetically as well as inertially confined fusion reactor. Plot these domains on an N vs. kT plane.

3.   Charged Particle Scattering

A common approach to the achievement and sustainment of fusion reactions involves conditions in which the fuel mixture exists in a plasma state. Such a state of rapidly moving ions and electrons provides for extensive scattering due to Coulomb force effects. These are particularly important because they lead to kinetic energy variations amongst particles, and more importantly, to particle losses from the reaction region thereby affecting the energy viability of the plasma.

3.1   Collisional Processes

Collisions between atomic, nuclear, and subnuclear particles take many forms. The important process of fusion between light nuclides represents a "discrete" inelastic process of nucleon rearrangement in which the reactants lose their former identity. In contrast, Coulomb scattering among ions and electrons causes "continuous" changes in direction of motion and kinetic energy. All these phenomena occur in a plasma to a varying extent and are therefore important in all confinement devices.

There may also exist a need to describe other selected collisional events in a fully or partially ionized medium, requiring therefore that the distinguishing characteristics of various processes be identified. Among these we note atomic processes such as photo-ionization, electron impact excitation, fluorescence, charge transfer and recombination, among others. Nuclear processes include inelastic nuclear excitation, nuclear de-excitation and elastic scattering.

A commonly occurring and important type of collision in a plasma is charged particle scattering attributable to the mutual electrostatic force. Such Coulomb scattering can vary from the most frequently occurring small-angle "glancing" encounters due to long-range interactions, up to the least likely near "head-on" collisions. The Coulomb scattering probability for ions is much larger than that to undergo fusion. Note that the deflections encountered in scattering reactions may lead to significant bremsstrahlung radiation power losses which lower the plasma temperature.

In general, the complete analysis of charged particle scattering is physically complex and mathematically tedious. As a consequence, we chose here to employ selected reductions in order to convey some of the essential and dominant features of specific relevance for our purposes here.

31

3.2   Differential Cross Section

Consider an isolated system of two ions of charge q_a and q_b, and possessing corresponding masses m_a and m_b. The magnitude of the Coulomb force is given by

$$F_c = \frac{1}{4\pi\,\varepsilon_o}\frac{q_a q_b}{r^2} \qquad (3.1)$$

where r is the distance of separation at any instant; evidently, as the charges move, this distance and also the direction of the force varies with time. The resultant trajectory of the particles a and b is suggested in Fig.3.1 for charged particle repulsion and attraction. In order to clearly specify a cross section for this process, it is useful to introduce the impact parameter r_o and the scattering angle θ_s with respect to the initial and final asymptotic particle trajectories. In Fig. 3.1 the two colliding particles are shown in antiparallel motion because such a simplified situation always applies if the collision is analyzed in the centre of mass reference frame. Thus, the deflection angle θ_s here indicated is identical to the scattering angle in the centre of mass system θ_c, which is associated with the directional change of the relative velocity v_r, to be subsequently introduced. Intuition based on physical grounds suggests a rigid relationship between the three parameters r_o, θ_c and v_r.

For the case of azimuthally symmetric scattering, Fig. 3.2, and which here applies because of the specific form of F_c, every ion of mass m_b and charge q_b moving through the ring of area

$$dA = 2\pi\,r_o dr_o \qquad (3.2)$$

will scatter off particle a—which has a charge of the same sign—through an angle θ_c into the conical solid angle element $d\Omega^*$ associated with the shadowed ring of Fig. 3.2; this conical solid angle element is evidently

$$d\Omega^* = 2\pi\sin(\theta_c)d\theta_c\;. \qquad (3.3)$$

The number of ions which scatter into a solid angle element can change substantially with r_o and v_r; it is therefore necessary to look for an angle dependent cross section $\sigma(\Omega)$ which here, due to azimuthal symmetry, is only a function of θ_c and yields a total scattering cross section σ_s according to

$$\sigma_s = \int_0^{4\pi}\sigma_s(\theta_c)d^2\Omega = \int_0^{\pi}\sigma_s(\theta_c)d\Omega^* \qquad (3.4)$$

implying the relation

$$\sigma_s(\theta_c) = \frac{d\sigma_s}{d\Omega^*}\;. \qquad (3.5)$$

That is, $\sigma_s(\theta_c)$ is a function of the conical solid angle Ω^* corresponding to a specified θ_c and possesses units of barns per steradian (b/sr). Since $d\sigma_s$ represents the differential cross sectional target area dA of Fig.3.2, and all ions entering this area will depart through the solid angle element dΩ*, we may write

$$N2\pi r_o dr_o = N\sigma_s(\theta_c)d\Omega^* \tag{3.6}$$

with N denoting a particle number per unit area, which, upon insertion of Eq.(3.3) yields specifically

$$\sigma_s(\theta_c) = \frac{r_o}{sin(\theta_c)}\left|\frac{dr_o}{d\theta_c}\right| \tag{3.7}$$

where $0 \le \theta_c \le \pi$. Here, the standard absolute-value notation for the Jacobian of a transformation has been incorporated. The process discussed above refers to the so-called Rutherford scattering and $\sigma_s(\theta_c)$ is known as the corresponding differential scattering cross section.

Case of Repulsion

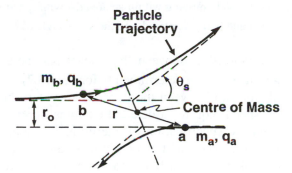

Case of Attraction

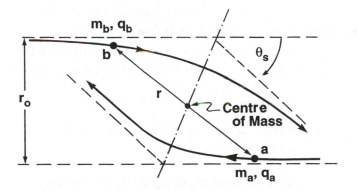

Fig. 3.1: Trajectories of ions "b" and "a" in the centre of mass reference frame under the influence of Coulomb repulsion and attraction.

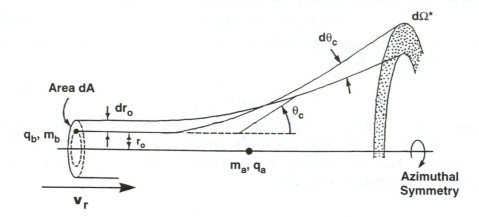

Fig. 3.2: Depiction of an ion b, moving within an area dA with speed v_r towards another ion a and being scattered into the solid angle $d\Omega*$.

The functional relationship between the impact parameter r_o and the polar angle of scattering θ_c needs to be determined before Eq.(3.7) can be specified.

The precise relationship between the impact parameter, the scattering angle, and the relative ion speed requires a detailed examination of the kinetics of the process together with the specification of energy and momentum conservation. These considerations and associated algebraic procedures are very cumbersome and will not be undertaken here. Suffice it to summarize that for the general case of arbitrary motion of two charged particles, the important relations can be compactly stated as

$$tan\left(\frac{\theta_c}{2}\right)=\left(\frac{q_a q_b}{4\pi \varepsilon_o}\right)\left(\frac{1}{m_r v_r^2 r_o}\right)$$

(3.8)

where the parameters introduced are as follows:

θ_c = scattering angle for particle b off of particle a in the COM (Centre of Mass) system which relates to θ_L in the LAB (Laboratory) system according to

$$cot(\theta_L)= \frac{m_b}{m_a} csc(\theta_c)+ cot(\theta_c) ;$$

(3.9a)

m_r = reduced mass of the two body system
$= (m_a m_b) / (m_a + m_b) ;$

(3.9b)

v_r = relative speed of the two particles of interest
$= \left| v_a - v_b \right| .$

(3.9c)

The meaning of the impact parameter r_o is unchanged.

Equation (3.8) provides a useful connection between the impact parameter r_o, the scattering angle θ_c in the COM system, and the relative particle speed v_r. For simplicity, we may write

$$tan\left(\frac{\theta_c}{2}\right) = \frac{K}{r_o} , \quad K = \frac{q_a q_b}{4\pi \, \varepsilon_o \, m_r \, v_r^2} \tag{3.10}$$

with K thus a parameter of the charge, mass, and the kinetic state of the two collision partners. To complete Eq. (3.7), we recall our previous comment on the geometry-of-collision of interest, Fig. 3.1, which allows the scattering angle θ_s introduced therein to be interchanged with θ_c, and use Eq. (3.10) to determine $dr_o/d\theta_c$ for K as a constant and thereby obtain the differential scattering cross section in the COM reference frame. That is, we get

$$\tfrac{1}{2} sec^2\left(\frac{\theta_c}{2}\right) d\theta_c = -\frac{K}{r_o^2} dr_o \tag{3.11}$$

and hence

$$\left|\frac{dr_o}{d\theta_c}\right| = \frac{r_o^2}{2K} sec^2\left(\frac{\theta_c}{2}\right). \tag{3.12}$$

Then, substitution of this expression in Eq.(3.7) gives

$$\sigma_s(\theta_c) = \left(\frac{r_o}{sin(\theta_c)}\right)\left(\frac{r_o^2 \, sec^2(\theta_c/2)}{2K}\right). \tag{3.13}$$

Now using Eq.(3.10) to eliminate the impact parameter r_o and employing the general half-angle identity

$$sin(\theta) = 2cos\left(\frac{\theta}{2}\right)sin\left(\frac{\theta}{2}\right) \tag{3.14}$$

yields, upon algebraic simplification, the final expression of relevance:

$$\sigma_s(\theta_c) = \frac{K^2}{4\,sin^4(\theta_c/2)} = \frac{1}{4}\left(\frac{q_a q_b}{4\pi \, \varepsilon_o \, m_r \, v_r^2}\right)^2 \frac{1}{sin^4(\theta_c/2)} \tag{3.15}$$

This explicit algebraic form is frequently called the Rutherford scattering cross section.

The more significant information about this charged particle interaction cross section is suggested in Fig. 3.3 and illustrates the following important result: the cross section approaches $\sigma_s(\theta_c) \rightarrow K^2/4$ for "head-on" collisions (i.e. $r_o \rightarrow 0$) and becomes arbitrarily large for increasingly smaller "glancing" angles (i.e. $r_o \rightarrow \infty$).

An informative conclusion about the singularity as $\theta_c \rightarrow 0$ is that, for example, if there were to exist only two nonstationary charged particles in the universe then some deflection would occur with certainty regardless of how far apart they were; that is, the Coulomb $1/r^2$ force dependence may become infinitesimally small at sufficient separations but, in principle, it never vanishes. Identifying a "reasonable" length beyond which charge effects can be considered to be unimportant, or even be ignored, is provided by the Debye length concept to be discussed next.

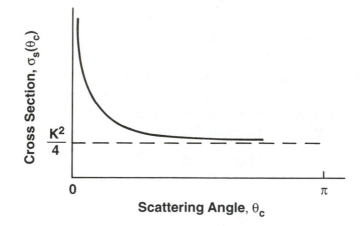

Fig. 3.3: Depiction of the Rutherford cross section as a function of scattering angle. Note
$$\sigma_s(\theta_c) \to \infty \text{ as } \theta_c \to 0.$$

3.3   Debye Length

Though a plasma is globally neutral, it may well acquire local charge variations which establish an electric potential and give rise to an electric field . The thermal motion of ions and electrons will therefore be influenced by the consequent force effects. An indication of the spatial extent of such an effect is represented by the Debye length. The following analysis yields a useful explicit expression for this important parameter.

Consider a dominant positive charge or electrode inserted in a plasma, Fig. 3.4. Due to the mobility of electrons, a negatively charged cloud will immediately form around this point with its density decreasing with distance. Similarly, a negative charge will also create a positive ion cloud spherically symmetric about it. Obviously, a plasma tends to shield itself from applied electric fields; that is, if an electrode is inserted into a plasma it will affect only its immediate surroundings. If the plasma were cold, one would observe as many charges in the surrounding cloud as are required to neutralize the inserted charge, Fig. 3.4. However, due to the finite plasma temperature, the plasma particles possess a substantial kinetic energy of thermal motion so that some–particularly those at the edge of the cloud–will escape from the shielding cloud by surmounting the electrostatic potential well which, as is known, decreases with increased distance r.

Evidently, we need to determine some characteristic shielding range and consider for that–over a small distance r from an inserted charge–some little perturbation of the electric charge density in the plasma. Beyond this range, a uniform neutral plasma continues to reside. The local electric field **E** thus established is related to the local charge density ρ^c by Maxwell's First Equation

$$\nabla \cdot \mathbf{E} = \frac{\rho^c}{\varepsilon_o} \ . \tag{3.16}$$

With this conservative field, a scalar potential function Φ is identified and related to the electric field by

$$\mathbf{E} = -\nabla \Phi \ . \tag{3.17}$$

By substitution in Eq.(3.16) and specializing for the case of interest, Fig. 3.4., Poisson's Equation takes on the form

$$\nabla \cdot (-\nabla \Phi) = \frac{\rho^c}{\varepsilon_o} \tag{3.18}$$

which, upon introducing the definition

$$\rho^c = \sum_{j=e,i} q_j N_j \ , \tag{3.19}$$

is written for a plasma containing only one species of ions as

$$-\nabla^2 \Phi = \frac{q_e N_e + q_i N_i}{\varepsilon_o} = \frac{-|q_e|}{\varepsilon_o}\left(N_e - N_i\right) \tag{3.20}$$

in which q_i and q_e are the ion and electron charge, and N_i and N_e are the local ion and electron densities, respectively. Determining the potential function Φ requires a knowledge of N_i and N_e as functions of position or of Φ.

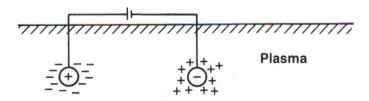

Fig. 3.4: Local charge variation in a plasma upon insertion of two dominant point charges.

The particle densities N_j, however, at thermodynamic equilibrium and in the presence of a potential energy Φ are known to depend upon Φ, the specific charge and the equilibrium temperature T_j according to the Boltzmann relation

$$N_e = C \exp\left(\frac{-q_e \Phi}{kT_e}\right) \tag{3.21a}$$

with the factors C_j being determined from the evident affinity $\Phi \to 0$ as $r \to \infty$ to represent the undisturbed background densities $N_j(r \to \infty)$. Equation (3.21a) may then be expanded by a Taylor series to give

$$N_e \approx N \left[1 + \frac{|q_e|\Phi}{kT_e} + \frac{1}{2} \left(\frac{|q_e|\Phi}{kT_e} \right)^2 \cdots \right]. \tag{3.21b}$$

This refers to the region where $| q_i\Phi / kT_i | \ll 1$, which is actually the dominant contributor to the thickness of the shielding cloud. Retaining only the linear terms of the expansion and substituting into Eq. (3.20) yields

$$-\nabla^2 \Phi \approx \frac{-|q_e|}{\varepsilon_o} \left[N \left(1 + \frac{|q_e|\Phi}{kT_e} \right) - N \right] = -\frac{|q_e|N}{\varepsilon_o} \left(\frac{|q_e|\Phi}{kT_e} \right) \tag{3.22}$$

where we have used the charge-neutrality boundary condition

$$q_e N_e (r \to \infty) + q_i N_i (r \to \infty) = 0. \tag{3.23}$$

The defining equation for the electrical potential function Φ, Eq. (3.22), is evidently of the form

$$\nabla^2 \Phi - \frac{\Phi}{\lambda_D^2} = 0 \tag{3.24}$$

with λ_D defined as the Debye length

$$\lambda_D = \sqrt{\frac{\varepsilon_o kT_e}{q_e^2 N}}. \tag{3.25}$$

Denoting the length for shielding ions by electrons by

$$\lambda_{De} = \sqrt{\frac{\varepsilon_o kT_e}{q_e^2 N_e (r \to \infty)}} \tag{3.26a}$$

and for screening electrons by ions by

$$\lambda_{Di} = \sqrt{\frac{\varepsilon_o kT_i}{q_i^2 N_i (r \to \infty)}} \tag{3.26b}$$

we realize the relation

$$\frac{1}{\lambda_D^2} = \frac{1}{\lambda_{De}^2} + \frac{1}{\lambda_{Di}^2}. \tag{3.27}$$

Taking $T_e = T_i = T$ and assuming the presence of singly-charged ions only, that is $N_i(r \to \infty) = N_e(r \to \infty) = N/2$, notably simplifies Eq. (3.25) to the expression

$$\lambda_D = \sqrt{\frac{\varepsilon_o kT}{q_e^2 N}}. \tag{3.28a}$$

The important dependence of λ_D is therefore

$$\lambda_D \propto \sqrt{\frac{T}{N}} \tag{3.28b}$$

with typical values of interest to thermonuclear fusion being in the range of about

1 μm to 1 mm.

The role of λ_D may be interpreted as a "shielding length" parameter for a plasma and becomes evident by solving Eq.(3.22) for the boundary conditions

$$\Phi(r=0)=\Phi_o \quad \text{and} \quad \Phi(r\to\infty)=0. \tag{3.29}$$

For the case of spherical geometry one obtains for the potential function in a plasma

$$\Phi_{plasma} \propto \frac{exp(-r/\lambda_D)}{r} \tag{3.30a}$$

which may be compared to the free-space potential given by

$$\Phi_{free\ space} \propto \frac{1}{r}. \tag{3.30b}$$

Thus, the potential Φ associated with the imposed electrostatic perturbation is attenuated in a plasma according to the magnitude of λ_D and is commonly said to be shielded to the distance of the Debye length.

This λ_D parameter is thus a useful concept and application of it relates to the elimination of the cross section singularity of Sec.3.2 as we will show next.

3.4 Scattering Limit

The singularity of the Coulomb scattering cross section, for $\theta_c \to 0$, Eq.(3.15), becomes particularly apparent in the evaluation of the total Coulomb scattering cross section. That is, substitution of Eq.(3.15) into Eq.(3.4) gives

$$\sigma_s = 2\pi \int_{\theta_c=0}^{\pi} \left(\frac{K^2}{4\ sin^4(\theta_c/2)}\right) sin(\theta_c)d\theta_c = \pi\ K^2 \int_{\theta_c=0}^{\pi} \frac{cos(\theta_c/2)}{sin^3(\theta_c/2)}d\theta_c. \tag{3.31}$$

To avoid the singularity as $\theta_c \to 0$, consider a finite non-zero lower bound θ_{min} and write

$$\sigma_s = 2\pi K^2 \int_{sin(\theta_{min}/2)}^{sin(\pi/2)} \frac{d[sin(\theta)]}{sin^3(\theta)}, \quad \theta = \frac{\theta_c}{2}$$

$$= \pi\ K^2 \left\{\left[sin\left(\frac{\theta_{min}}{2}\right)\right]^{-2} - 1\right\}. \tag{3.32}$$

Note in this context that it is the collisions with small deflections which are responsible for the total Coulomb scattering cross section σ_s becoming very large.

A number of physical considerations suggest that θ_{min} should be chosen in accordance with a maximum impact parameter beyond which Coulomb scattering is relatively small. In a vacuum, the Coulomb field from a dominant isolated charge extends to infinity, implying therefore a scattering "deflection" interaction at any arbitrary impact parameter. In a plasma, however, the target charge would be surrounded by a particle cloud of opposite charge. The fields due to surrounding particles will effectively "screen-out" the Coulomb field of an arbitrary charge at some maximum impact parameter distance which corresponds to some minimum scattering angle. This distance may be specified as the radius of an imaginary sphere surrounding a target ion such that the plasma electrons will reduce the target ion's Coulombic field by $1/e$ at the sphere's surface. The Debye length, λ_D as defined by Eq.(3.23a), evidently corresponds to this distance and we take the maximum impact parameter as

$$r_{o,max} = \lambda_D .$$ (3.33)

Thus, θ_{min} follows from the inversion of Eq.(3.10):

$$\theta_{min} = 2 \tan^{-1}\left(\frac{K}{\lambda_D}\right).$$ (3.34)

Figure 3.5 displays the Coulomb scattering cross section for the case of d-t interactions and for comparison also shows the fusion cross section; as suggested previously, it is evident that Coulomb scattering events occur orders of magnitude more frequently than fusion reactions.

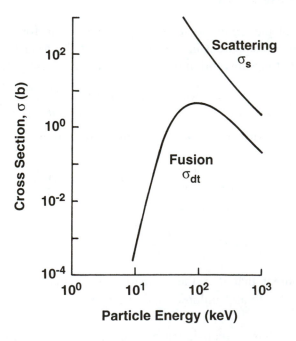

Fig. 3.5: Scattering cross section and fusion cross section for deuterium incident on a tritium target.

3.5 Bremsstrahlung Radiation

An important consequence of scattering in fusion plasmas is bremsstrahlung radiation, which refers to the process of radiation emission when a charged particle accelerates or decelerates. It involves the transformation of some particle kinetic energy into radiation energy which–due to its relatively high frequency (X-ray wavelength range of ~ 10^{-9} m)–may readily escape from a plasma; thus, the kinetic energy of plasma particles is reduced, plasma cooling occurs, and a compensating energy supply may be required in order to maintain the desired plasma temperature.

A rigorous derivation of bremsstrahlung power emission in a hot plasma involves considerations of quantum mechanics and relativistic effects and is both tedious and time consuming. Indeed, even advanced methods of analysis require the imposition of simplifying assumptions if the formulation is to be at all tractable. It is, however, less difficult to develop an approximation of the dominant effect of the bremsstrahlung processes by the following considerations. A particle of charge q_e and moving with a time varying velocity $\mathbf{v}(t)$ will– according to classical electromagnetic theory in the nonrelativistic limit–emit radiation at a power

$$P \propto q_e^2 \left| \frac{d\mathbf{v}}{dt} \right|^2 .$$

(3.35)

In Fig. 3.6, we suggest this process for an electron moving in the electrostatic field of a heavy ion. To estimate the energy radiated per encounter, we replace $|d\mathbf{v}/dt|$ in Eq.(3.35) by an average acceleration $\bar{a}$ to be calculated from Newton's Law, $\bar{a} = \overline{F}_c / m$, with $\overline{F}_c$ representing the magnitude of the average Coulomb force approximated here by the electrostatic force between the two interacting particles when separated by the impact parameter r_o. That is, we take

$$\bar{a} \approx \frac{F_c(r_o)}{m} = \frac{Z q_e^2}{4\pi \varepsilon_o r_o^2 m}$$

(3.36)

for the average acceleration experienced by a particle of mass m and charge q_e in the field of a charge $q_i = -Zq_e$. Since an electron possesses a mass 1/1836 of that of a proton–with an even smaller ratio existing when compared to a deuteron or triton–the electrons in a plasma will therefore be the main contributors to bremsstrahlung radiation. We suggest this in Fig. 3.7 for a representative electron trajectory undergoing significant directional changes in a background of sluggish ions.

Imagining an electron (m_e, q_e) to be in the vicinity of an ion (m_i, $|q_i| = |Zq_e|$) for a time interval

$$\Delta t_o \approx \frac{r_o}{v_r}$$

(3.37)

with $\overline{v}_r$ denoting the average relative speed between the electron and ion, we combine Eqs.(3.35) to (3.37) to assess the bremsstrahlung radiation energy per collision event as

$$\frac{E_{br}}{\text{collision}} \approx P\Delta t_o \propto \frac{Z^2 q_e^6}{m_e^2 r_o^3 \overline{v}_r}. \tag{3.38}$$

Note this expression refers to the collision of a single electron with an ion at the specific impact parameter r_o. Extending these considerations to a bulk of electrons of density N_e all approaching the ion with $\overline{v}_r$, then the number of electrons colliding per unit time with the same ion at the same impact parameter r_o is given by

$$\frac{dR_o}{\text{ion}} \approx N_e \overline{v}_r 2\pi r_o dr_o. \tag{3.39}$$

Multiplying this relation by the ion density N_i we obtain the differential electron-ion collision rate density

$$dR_o \approx N_e N_i \overline{v}_r 2\pi r_o dr_o \tag{3.40}$$

thereby accounting for all electron-ion interactions per unit volume and per unit time occurring about a specific impact parameter r_o.

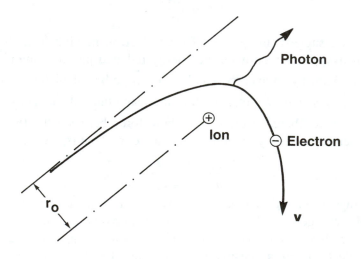

Fig. 3.6: Depiction of photon emission from an accelerated electron passing near an ion.

Next, in order to obtain the total bremsstrahlung radiation power density associated with the entire electron-ion force impact area, Eq. (3.38) needs to be multiplied by Eq.(3.40) and integrated over the range of the impact parameter $r_{o,min}$ to $r_{o,max}$; that is we now obtain for the bremsstrahlung power the proportionality

$$P_{br} \propto N_i \, N_e \frac{Z^2 \, q_e^6}{m_e^2} \int_{r_{o,min}}^{r_{o,max}} \frac{dr_o}{r_o^2}.$$ (3.41)

While $r_{o,max} = \infty$ can be imposed, a finite minimum impact parameter needs to be specified due to the integral singularity for $r_o \rightarrow 0$. We choose to identify $r_{o,min}$ with the DeBroglie wavelength of an electron, that is,

$$r_{o,min} = \frac{h}{m_e \, \overline{v_e}}$$ (3.42)

where h is Planck's constant and $\overline{v_e}$ represents the average thermal electron speed defined by Eq.(2.19b):

$$\overline{v_e} = \sqrt{\frac{8 \, kT_e}{m_e \pi}}.$$ (3.43)

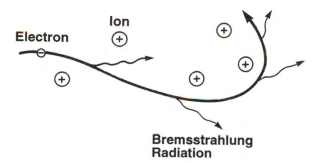

Fig. 3.7: Schematic depiction of an electron trajectory and bremsstrahlung radiation.

With the integration limits thus specified, the integral of Eq.(3.41) is readily evaluated so that the electron bremsstrahlung radiation power density is found to exhibit the following dominant dependencies:

$$P_{br} = A_{br} \, N_i \, N_e \, Z^2 \, \sqrt{kT}$$ (3.44)

where A_{br} is a constant of proportionality. For kT in units of eV and particle densities in units of m^{-3},

$$A_{br} \approx 1.6 \times 10^{-38} \left(\frac{m^3 \, J}{\sqrt{eV} \, s} \right)$$ (3.45)

which yields P_{br} in units of $W \cdot m^{-3}$. An important point to note here is that P_{br} is proportional to $(kT)^{1/2}$.

Problems

3.1   Examine the relationship between θ_C and θ_L for the limiting cases of $m_b/m_a = \{0, 1, \infty\}$.

3.2   For a typical case of d-t interaction at 5 keV, plot $\sigma_s(\theta_c)$, $0 \le \theta_c \le \pi$.

3.3   For the conditions in problem 3.2, calculate σ_s assuming a background particle density of 10^{20} m^{-3}.

3.4   Calculate the bremsstrahlung power increase if a d-t plasma were to contain totally stripped oxygen ions at a concentration of 1% of the electron density.

3.5   Calculate the ratio of bremsstrahlung power to fusion power for a d-t plasma with $N_i = N_e = 10^{20}$ m^{-3} at 2 keV and 20 keV.

3.6   Determine from a plot of power density versus kinetic temperature (use a double logarithmic scale) for a 50:50% D-T fusion plasma
      (a) the so called ideal d-t ignition temperature T^*, i.e. the temperature at which the plasma fusion power density equals the loss power density due to bremsstrahlung.
      (b) the temperature T^*_{ign} more relevant to fusion ignition since it refers to the plasma operational state where the bremsstrahlung loss power is balanced by the energy transfer from the fusion alphas to the background plasma per unit time by Coulomb collisions. Assuming that the charged particle fusion power is entirely transferred, T^*_{ign} is found from

$$f_{c,dt} \, P_{dt}\left(T^*_{ign}\right) = P_{br}\left(T^*_{ign}\right)$$

with $f_{c,dt}$ representing the fraction of d-t fusion power allocated to charged particles (in this case α's). What does this condition mean for the overall power balance of the fusion plasma if other heat losses were neglected?

3.7   A fusion reactor using two opposing accelerators is proposed, where a 30 keV tritium beam from one is aimed head on at a 30 keV deuterium beam from the other. Would this work, and can you suggest improvements? What would you estimate is the maximum energy gain possible with this system (see Ch. 8)?

PART II   CONFINEMENT, TRANSPORT, BURN

4. Fusion Confinement

The attainment of a sufficiently high reaction-driven energy density is a requirement of all energy systems. For fusion it is essential that the reactant nuclei attain a sufficiently high kinetic energy of relative motion in order to achieve substantial rates of exothermic reactions. These conditions must then be retained for a sufficiently long time in a specified reaction domain. Confining the interacting fuel particles at an appropriate high temperature is thus a most basic consideration of fusion energy systems.

4.1   Necessity of Confinement

Unlike fission reactions which involve a neutral reactant and thus do not experience repulsive effects, fusion reactants are positively charged and must overcome their electrostatic repulsion in order to get close enough for the strong nuclear forces of attraction to dominate. Hence, the essential condition for fusion is the requirement for a sufficiently high kinetic temperature of the reacting species in order to facilitate the penetration of the Coulomb barrier.

The attainment of ion energies in excess of this Coulomb barrier, which is about 370 keV for d-t fusion, poses little technical difficulty. For example, readily available medium-energy accelerators could be used to inject deuterons, of say $E_d \approx 500$ keV, into a tritiated target; surrounding neutron and alpha detectors could then be used to identify the reaction products as evidence of whether the reaction $d + t \rightarrow n + \alpha + 17.6$ MeV had taken place. Obviously, if each injected deuteron were to lead to d-t fusion, then the energy multiplication would be $E_{out} / E_{in} = 17.6 / 0.5 = 35$ and thus adequate for fusion energy utility purposes.

Theory suggests and experiment has confirmed that such a beam-target concept is totally inadequate for the following reasons: as beam deuterons enter a target they lose energy through the processes of ionization and heating the target. As discussed in the preceding chapter and displayed in Fig. 3.5, they are far more likely to scatter–rather than fuse–with an additional attendant energy loss by bremsstrahlung radiation. Thus, very quickly, the projectiles will have slowed down to energies far below the Coulomb barrier rendering further fusion reactions most unlikely. Thus, the overall fusion energy release can not exceed the energy required for beam acceleration. The futility of this approach was recognized early in fusion energy research.

A more promising approach, however, soon emerged. One begins with a population of deuterium and tritium atoms in some confined space, and by heating one causes both ionization and the attainment of high temperature of the fuel ions. The resulting ensemble of positive and negative charges thus forms a plasma which is expected to attain thermodynamic equilibrium as a result of random collisions. The resultant spectrum of particle energies is then well described by a Maxwell-Boltzmann distribution with the high energy part of this distribution providing for most of the desired fusion reactions. Because the reaction activation occurs here due to random thermal motion of the reacting nuclei, this process is therefore called thermonuclear fusion. The critical technical requirement is the sustainment of a sufficiently stable high temperature ($\sim 10^8$ K) plasma in a practical reaction volume and for a sufficiently long period of time to render the entire process energetically viable. Confinement of the fuel ions by some means is thus crucial to maintain these conditions within the required reaction volume.

We add that in contrast to this high-temperature approach to fusion, there exists also a low-temperature approach free of the above type of confinement problems. As we will show in Ch. 12, confinement by atomic-molecular effects may also be exploited.

4.2   Material Confinement

The simplest and most obvious method with which to provide confinement of a plasma is by a direct-contact with material walls, but is impossible for two fundamental reasons: the wall would cool the plasma and most wall materials would melt. We recall that the fusion plasma here requires a temperature of $\sim 10^8$ K while metals generally melt at a temperature below 5000 K.

Further, even for a plasma not in direct contact, there exist problems with a material wall. High temperature particles escaping from the plasma may strike the wall causing so-called "sputtered" wall atoms to enter the plasma. These particles will quickly become ionized by collisions with the background plasma and can appear as multiply-charged ions which are known as fusion plasma impurities. Then, as shown in the analysis leading to Eq. (3.44), the bremsstrahlung power losses increase with Z^2 thus further cooling the plasma.

4.3   Gravitational Confinement

A most spectacular display of fusion energy is associated with stars, where confinement comes about because of the gravitational pressure of an enormous mass. High density and temperature thereby result toward the stellar centre enabling the fusile ions to burn. While energy leakage and particle escape occurs from the star's surface, the interior retains most of the reaction power and prevails against the occurrent radiation pressure through the deep gravitational potential wells, thus assuring stable confinement for times long enough to burn most of the stellar fusionable inventory.

Since fusion-powered stars possess dimensions and masses of such enormity , it is evident that confinement by gravity cannot be attained in our terrestrial environment.

4.4   Electrostatic Confinement

A more plausible approach to the confinement problem is to recognize that ions are known to be affected by electrostatic fields, so that confinement by such force effects can also be conceived.

One specific and interesting electrostatic approach to confinement involves a spherical metallic anode emitting deuterons towards the centre. These ions then pass through a spherical negatively charged grid designed to be largely transparent to these deuterons. The positive ions converge toward the centre and a positive space charge forms tending to reverse the motion of the ions and thereby establish a positive ion shell inside the hollow cathode. This internal positive ion shell is called a "virtual anode" in order to distinguish it from the "real anode" which produced these ions.

The metallic cathode, situated between the real and virtual anode, also emits electrons towards the centre. After passage through the virtual anode, the electrons similarly form a "virtual cathode" located further towards the centre. It is conceived that several such nested virtual cathodes and virtual anodes will form with the ion density increasing toward the centre. Finally, then, fusion reactions are expected to occur in these inner ion shells of increasing particle density.

However, in general, the electric fields required, the likelihood of discharge breakdown, and problems of geometrical restrictions have generally served to limit consideration of electrostatic confinement in the pursuit of fusion reaction rates suitable for an energetically viable system.

4.5   Inertial Confinement

A confinement method with apparently more merit than the aforementioned is inertial confinement fusion, involving compression of a small fusion fuel pellet to high density and temperature by external laser or ion beams, Fig.4.1. The density-temperature conditions so achieved are expected to provide for a pulse of fusion energy before pellet disassembly. The incident laser or ion beam induces an inward directed momentum of the outer layers of the pellet, thereby yielding a high density of material, the inertia of which confines the fuel against the fusion reaction explosive effect of disassembly for a sufficient time to allow enough fusion reactions to occur.

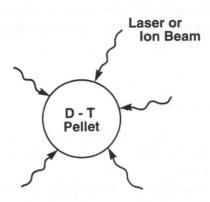

Fig. 4.1: Compression and heating of a d-t pellet by external laser or ion beams.

Some critical characteristic features of inertially confined fusion can be described by the following. Consider, for this purpose, a d-t pellet at an advanced stage of compression with a plasma formed therefrom and fusion burn occurring. The tritium ion density N_t will, in the absence of leakage, decrease according to its burn rate

$$\frac{dN_t}{dt} = - <\sigma v>_{dt} N_d N_t \tag{4.1}$$

with the symbols here used as previously defined. In this context, we may further identify a triton mean-life, τ_t, by

$$\frac{dN_t}{dt} = - \frac{N_t}{\tau_t} . \tag{4.2}$$

This triton mean-life may evidently be identified as the mean time between fusion events, τ_{fu}, which by equating Eq.(4.1) with (4.2), yields specifically

$$\tau_{fu} \approx \frac{1}{<\sigma v>_{dt} N_{d,o}} \tag{4.3}$$

with $N_d = N_{d,o}$ taken to be some suitable initial average deuterium population density and $<\sigma v>_{dt}$ is taken at some appropriate average temperature. Expecting the pellet disassembly to be rapid suggests that τ_{fu} should also be small. Two possibilities of reducing τ_{fu} become evident from Eq.(4.3). One can enhance $<\sigma v>$ by increasing the relative speed of the reactants which, however, is limited by the techniques of heat deposition in the fuel as well as by the more rapid disintegration at higher temperatures. The other option is to increase the fuel density by several orders of magnitude. The reaction rate parameter $<\sigma v>_{dt}$ is a maximum at an ion temperature of ~60 keV, and the density is a maximum at the onset of fusion burn which is also occurring at the time pellet disassembly begins.

A pellet, once compressed and with fusion reactions taking place, will evidently heat up further and hence tend toward disintegration. The speed of outward motion of the pellet atoms is, to a first approximation, given by the sonic speed v_s which, in a d-t plasma, is given by

$$v_s = \sqrt{\gamma \frac{2N_i kT}{N_i m_i}} = \sqrt{\frac{10}{3} \frac{kT_i}{m_i}}. \tag{4.4}$$

Here, γ is the ratio of specific heats ($\gamma = 5/3$), N_i is the ion density, and $\overline{m_i}$ the average ion mass.

As a characteristic expansion, we take the pellet's spherical inflation from its initial radius R_b when the fusion burn began, to a size of radius $2R_b$. During this process, the fuel density will have decreased by a factor of 2^3 and the rate of fusion energy release will have accordingly decreased by the factor $(2^3)^2 = 64$; hence, most of the fusion reactions will have taken place during this initial stage of disassembly. Thus, the time for doubling of the pellet radius is taken as being representative of the inertial confinement time, τ_{ic}, given by

$$\tau_{ic} = \frac{2R_b - R_b}{v_s} = R_b \left(\frac{3\overline{m_i}}{10kT_i} \right)^{1/2}. \tag{4.5}$$

Evidently, τ_{fu} of Eq.(4.3) should be shorter than–or perhaps of the order of–τ_{ic} for sufficient fusion burn to take place so that, as a required initial condition, we must have

$$\tau_{fu} < \tau_{ic}, \tag{4.6a}$$

that is,

$$\frac{1}{<\sigma v>_{dt} N_{d,o}} < R_b \left(\frac{3\overline{m_i}}{10kT_i} \right)^{1/2}. \tag{4.6b}$$

For the case of $N_{d,o} = N_{t,o} = N_{i,o}/2$, that is half of the compressed fuel density at the beginning of the fusion burn, we may therefore write the requirement as

$$N_{i,o} R_b > \frac{\left(\frac{40kT_i}{3\overline{m_i}} \right)^{1/2}}{<\sigma v>_{dt}}. \tag{4.7}$$

Taking an average fusion fuel temperature of about $E_{th} \approx 20$ keV, we obtain by substitution,

$$N_{i,o} R_b > 10^{24} \ \text{cm}^{-2}. \tag{4.8}$$

Some essential technical features of inertial confinement fusion may now be qualitatively and quantitatively established. First, as will be shown in Ch. 11, the beam energy required to compress a pellet corresponds to the resultant heat content of the compressed pellet, and hence varies as R_b^3. Existing laser beam powers are such that R_b needs to be kept in the range of millimetres or less. Second, the fuel density will evidently need to become very large, typically exceeding 10^3 times that of its equivalent solid density. Finally, the quantity of the initial fuel which burns up needs to be carefully specified since it relates to the overall energy viability of each fusion pulse as well as to the capacity of the surrounding medium to absorb the blast energy.

Thus, as an initial conclusion, we may assert that very high power drivers and very high compression is required for fusion energy achievement by inertial confinement. Further analysis of inertial confinement fusion is the subject of Ch. 11.

4.6   Magnetic Confinement

One of the most effective means of plasma confinement to date involves the use of magnetic fields. A particle of charge q and mass m located in a magnetic field of local flux density **B** is constrained to move according to the Lorentz force given by

$$m\frac{d\mathbf{v}}{dt} = q(\mathbf{v} \times \mathbf{B})$$
(4.9)

with **v** the velocity vector.

For the case of a solenoidal **B**-field produced by a helical electrical current with density **j**, Fig. 4.2a, ions and electrons will move (as will be shown in Ch. 5)–depending upon the initial particle velocity–either parallel or antiparallel to the **B**-field lines and spiral about them with a radius of gyro-motion of

$$r_g \propto \frac{1}{|\mathbf{B}|} .$$
(4.10)

Scattering reactions may, however, transport them out of these uniform spiral orbits with two consequences: they may be captured into another spiral orbit or they may scatter out of the magnetic field domain. In the absence of scattering they are essentially confined as far as directions perpendicular to **B** are concerned, traveling helically along **B** with an unaffected velocity component parallel to **B**. Subsequently, they will eventually depart from the region of interest.

Solenoidal fields belong to the oldest and most widely used magnetic confinement devices used in plasma physics research. The ion and electron densities may clearly be enhanced by increasing the magnetic field thereby also providing for a smaller radius of gyration. There exists, however, a dominant property which renders their use for fusion energy purposes most detrimental: for solenoidal dimensions and magnetic fields generally achievable, fuel ion leakage through the ends is so great that these devices provide little prospect for use as fusion reactors.

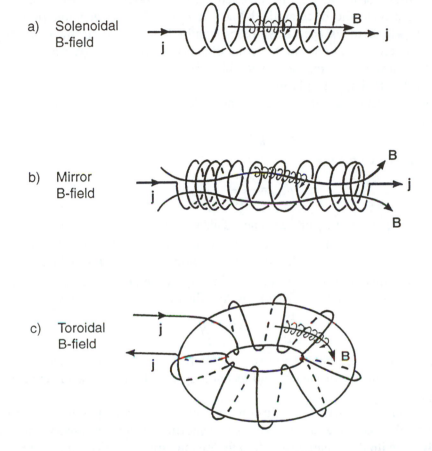

a) Solenoidal B-field

b) Mirror B-field

c) Toroidal B-field

Fig. 4.2: Depiction of three magnetic field topologies and illustrative ion trajectories.

If, however, the magnetic field strength is increased specifically at each end of the cylindrical region, i.e. the **B**-field lines appear to be substantially squeezed together at the ends, Fig. 4.2b, the number of leaking particles is considerably reduced. Such a squeezed field configuration is referred to as a magnetic mirror since it is able to reflect charged particles, as will be shown in Sec. 9.3. We note that ions and electrons possessing excessive motion along the magnetic axis will still penetrate the magnetic mirror throat.

The attractiveness of the mirror concept not withstanding, a magnetic mirror can thus not provide complete confinement and–as in all open-ended configurations–is associated with unacceptably high particle losses through the ends. Hence, to avoid end-leakage entirely, the obvious solution is to eliminate the ends by turning a solenoidal field into a toroidal field, Fig. 4.2c. The resultant toroidal magnetic field topology has spawned several important fusion reactor concepts; the most widely pursued of such devices is known as the tokamak, which will be discussed in Ch. 10.

At first consideration, the charged particles could be viewed as simply spiraling around the circular field lines in Fig. 4.2c, not encountering an end through which to escape. Any losses would have to occur by scattering or diffusion across field lines in the radial direction causing leakage across the outer surface. Further, we mention that collective particle oscillations may occur, thereby destabilizing the plasma.

One important plasma confinement indicator is the ratio of kinetic particle pressure

$$p_{kin} = N_i kT_i + N_e kT_e \qquad (4.11a)$$

to the magnetic pressure

$$p_{mag} = \frac{B^2}{2\mu_o} \qquad (4.11b)$$

with μ_o the permeability of free space. This ratio is defined as the beta parameter, β, and is a measure of how effectively the magnetic field constrains the thermal motion of the plasma particles. A high beta would be most desirable but it is also known that there exists a system-specific β_{max} at which plasma oscillations start to destroy the confinement. That is, for confinement purposes, we require

$$\beta_{max} \frac{B^2}{2\mu_o} \geq N_i kT_i + N_e kT_e . \qquad (4.12)$$

Thus, the maximum plasma pressure is determined by available magnetic fields thereby introducing magnetic field technology as a limit on plasma confinement in toroids.

In order to assess the fusion energy production possible in such a magnetically confined, pressure-limited deuterium-tritium plasma, we introduce Eq.(4.12), with the equality sign, into the fusion power density expression

$$P_{fu} = N_d N_t <\sigma v>_{dt} Q_{dt} \qquad (4.13)$$

to determine–for the case of $N_d = N_t = N_i/2$, $N_i = N_e$ and $T_i = T_e$ the magnetic pressure-limited fusion power density

$$P_{fu,mag} = \frac{\beta_{max}^2 B^4}{64 \mu_o^2} \cdot \frac{<\sigma v>_{dt}}{T_i^2} Q_{dt} \qquad (4.14)$$

which is displayed in Fig. 4.3 as a function of the plasma temperature.

Another overriding consideration of plasma confinement relates to the ratio of total energy supplied to bring about fusion reactions and the total energy generated from fusion reactions. Commercial viability demands

$$E_{fu} >> E_{supply} . \qquad (4.15)$$

Most toroidal confinement systems currently of interest are expected to operate in a pulsed mode characterized by a burn time τ_b. Assuming the fusion power P_{fu} to be constant during τ_b, or to represent the average power over that interval, the fusion energy generated during this time is given by

$$E_{fu} = P_{fu} \cdot \tau_b = <\sigma v>_{dt} N_d N_t Q_{dt} \tau_b \qquad (4.16)$$

for d-t fusion. For perfect coupling of the energy supplied to an ensemble of deuterons, tritons, and electrons, we have

$$E_{supply} = \tfrac{3}{2} N_d kT_d + \tfrac{3}{2} N_t kT_t + \tfrac{3}{2} N_e kT_e . \qquad (4.17)$$

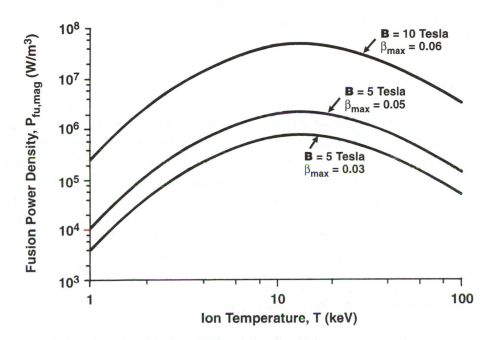

Fig. 4.3: Pressure-limited fusion power density in a magnetically confined d-t fusion plasma.

One may therefore identify an ideal energy breakeven of $E_{fu} = E_{supply}$ as defined by

$$<\sigma v>_{dt} N_d N_t Q_{dt} \tau_b = \tfrac{3}{2} N_d kT_d + \tfrac{3}{2} N_t kT_t + \tfrac{3}{2} N_e kT_e \qquad (4.18)$$

Imposing $N_d = N_t = N_i/2$, $N_i = N_e$, and $T_d = T_t = T_e = T$ we may simplify the relation to write for an ideal breakeven condition:

$$N_i \tau_b = \frac{12kT}{<\sigma v>_{dt} Q_{dt}} . \qquad (4.19)$$

Since $Q_{dt} = 17.6$ MeV and $<\sigma v>_{dt}$ is known as a function of kT, we readily compute the product

$$N_i \tau_b \sim 10^{20} \text{ s·m}^{-3} \qquad (4.20)$$

for $kT \approx 12$ keV. Note that for fusion burn to be assured over the period τ_b, it is required that the actual magnetic confinement time, $\tau_{mc} > \tau_b$. The incorporation of other energy/power losses as well as energy conversion efficiencies suggests that this ideal breakeven is an absolute minimum, and operational pulsed fusion systems need to possess a product much higher. For instance, the accelerated motion of charged particles inevitably is accompanied by electro-magnetic radiation emission, as described by Eq.(3.39), which constitutes power loss from a confined plasma.

There exist various versions of Eq. (4.19)–depending upon what energy/power terms and conversion processes are included; collectively they are known as Lawson criteria in recognition of John Lawson who first published such analyses.

Problems

4.1   Verify the expression for the sonic speed in a compressed homogeneous deuterium-tritium ICF pellet, given in Eq. (4.4), by evaluating the defining equation $v_s^2 = dp/d\rho$ with ρ denoting the mass density of the medium in which the sonic wave is propagating. Assuming the sound propagation is rapid so that heat transfer does not occur, the adiabatic equation of state, $p\rho^{-\gamma} = C$, is applicable to the variations of pressure and density. Determine the constant C using the ideal gas law. To quantify γ, which represents the ratio of specific heats at fixed pressure and volume, respectively, c_p/c_v, use the relations $c_p - c_v = R$ (on a per mole basis) and $c_v = fR/2$ known from statistical thermodynamics, where R is the gas constant and f designates the number of degrees of freedom of motion ($f = 3$ for a monatomic gas, $f = 5$ for a diatomic gas, ...).

4.2   Compute v_s of Eq.(4.4) for typical hydrogen plasmas and compare to sound propagation in other media.

4.3   Re-examine the specification of a mean-time between fusions, Sec. 4.5, for the more general case of steady-state fusion reactions. How would you define, by analogy, the mean-free-path between fusions, λ_{fu}?

4.4   If 10% burnup of deuterium and tritium is achieved in a compressed sphere with $R_b = 0.25$ mm and fulfilling the burn condition of Eq. (4.8) with $N_{i,o}R_b = 5 \times 10^{-28}$ m^{-2}, at what rate would these pellets have to be injected into the reactor chamber of a 5.5 GW$_t$ laser fusion power plant? The pellets contain deuterium and tritium in equal proportions. Power contributions from neutron-induced side reactions such as in lithium are to be ignored here.

4.5   Assess the minimum magnetic field strength required to confine a plasma having $N_i = N_e = 10^{20}$ m^{-3} and $T_i = 0.9\, T_e = 18$ keV, when $\beta_{max} = 8\%$.

4.6   Compute the maximum fusion power density of a magnetically confined d-h fusion plasma ($N_d = N_h$, $T_d = T_h = T_e$, no impurities) as limited by an assumed upper value of attainable field strength , B = 15 Tesla, dependent on the plasma temperature. Superimpose the result for $\beta_{max} = 2\%$ in Fig. 4.3.

5.   Individual Charge Trajectories

The need for plasma confinement requires that some control on ion and electron motion be considered. While moving charges accordingly generate electromagnetic fields which additionally act on other moving particles, for the case of low densities, these induced fields–as well as the effect of collisions–may be neglected so that the trajectories of charged particles may be considered to be governed entirely by external field forces acting on them.

5.1   Equation of Motion

The basic relation which determines the motion of an individual charged particle of mass m and charge q in a combined electric and magnetic field is the equation

$$m\frac{d\mathbf{v}}{dt} = q\mathbf{E} + q(\mathbf{v} \times \mathbf{B}).$$ (5.1)

Here, $\mathbf{v}$ is the velocity vector of the particle at an arbitrary point in space and $\mathbf{E}$ and $\mathbf{B}$ are, respectively, the local electric field and magnetic flux density perceived by the particle; the magnetic flux density is commonly called the magnetic field. The last term in this equation is also known as the Lorentz force. We assume SI units so that the units of $\mathbf{E}$ are Newton/Coulomb while for $\mathbf{B}$ the units are Tesla; note that 1 Tesla (T) is defined as 1 Weber·m^{-2} (= 10^4 Gauss = 1 kg·s^{-2}·A^{-1}).

While gravitational forces are of importance in stellar fusion, they are negligible compared to the externally generated magnetic field forces and the established electric fields associated with fusion energy devices of common interest; hence, gravitational force effects need not be included for our purposes.

5.2   Homogeneous Electric Field

As an initial case, consider a domain in space for which $\mathbf{B} = \mathbf{0}$ with only a constant electric field affecting the particle trajectory. Orienting the Cartesian coordinate system so that the z-axis points in the direction of $\mathbf{E}$, Fig. 5.1, and imposing an arbitrary initial particle velocity at t = 0, referenced to the origin of the coordinate system, we simply write for Eq.(5.1)

$$m\frac{d\mathbf{v}}{dt} = q\,E_z\mathbf{k}\,, \tag{5.2}$$

where $\mathbf{k}$ is the unit vector in the z-direction. This vector representation for the motion of a single charged particle in an electric field can be decomposed into its constituent components

$$\frac{dv_x}{dt} = 0\,, \quad \frac{dv_y}{dt} = 0\,, \quad \frac{dv_z}{dt} = \left(\frac{q}{m}\right)E_z\,, \tag{5.3}$$

and these can be solved by inspection:

$$v_x = v_{x,o}\,, \quad v_y = v_{y,o}\,, \quad v_z = v_{z,o} + \left(\frac{q}{m}E_z\right)t\,. \tag{5.4}$$

Integrating again with respect to time gives the position of the charged particle at any time t:

$$x = v_{x,o}t\,, \quad y = v_{y,o}t\,, \quad z = v_{z,o}t + \tfrac{1}{2}\left(\frac{q}{m}E_z\right)t^2\,. \tag{5.5}$$

The corresponding trajectory for a positive ion is suggested in Fig. 5.2 with the particle located at the origin of the coordinate system at t = 0. This description therefore constitutes a parametric representation of a curve in Cartesian (x,y,z) space and corresponds to the trajectory of the charged particle under the action of an electric field only and for the initial condition specified.

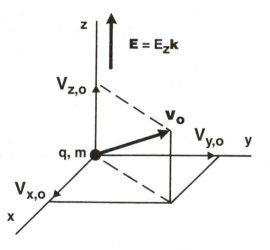

Fig. 5.1: Orientation of an electric field $\mathbf{E}$ acting on a particle of charge q and mass m.

The important feature to note is that the components of motion for this individual charged particle perpendicular to the $\mathbf{E}$-field, that is v_x and v_y, do not change with time; however, the velocity component in the direction of the $\mathbf{E}$-field, $v_z(t)$, is seen from Eq.(5.4) to linearly vary with time. The particle is accelerated in the direction of $\mathbf{E}$ for a positive charge and in the opposite direction for a negative charge.

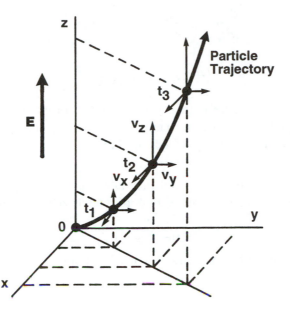

Fig. 5.2: Trajectory of a positively charged ion in a homogeneous electric field.

5.3   Homogeneous Magnetic Field

Consider now a single particle trajectory with $\mathbf{E} = \mathbf{0}$ and $\mathbf{B}$ constant in the space of interest. The relevant governing equation now becomes

$$m\frac{d\mathbf{v}}{dt} = q(\mathbf{v} \times \mathbf{B})\tag{5.6}$$

and by the definition of a vector cross product, the force on the particle is here perpendicular to both $\mathbf{v}$ and $\mathbf{B}$. Writing Eq.(5.6) in terms of its vector components, we find

$$\frac{d}{dt}\left(v_x\mathbf{i} + v_y\mathbf{j} + v_z\mathbf{k}\right) = \frac{q}{m}\left[\left(v_x\mathbf{i} + v_y\mathbf{j} + v_z\mathbf{k}\right) \times \left(B_x\mathbf{i} + B_y\mathbf{j} + B_z\mathbf{k}\right)\right]\tag{5.7}$$

where $\mathbf{i}$, $\mathbf{j}$ and $\mathbf{k}$ are orthogonal unit vectors along the x, y and z axes, respectively.

Orienting this coordinate system so that the z-axis is parallel to the $\mathbf{B}$-field, thereby imposing $B_x = B_y = 0$, Fig. 5.3, and equating the appropriate vector components of Eq.(5.7), we obtain the following set of differential equations together with their initial conditions:

$$\frac{dv_x}{dt} = \left(\frac{q\,B_z}{m}\right) v_y \,, \qquad v_x(0) = v_{x,o} \,, \tag{5.8a}$$

$$\frac{dv_y}{dt} = -\left(\frac{q\,B_z}{m}\right) v_x \,, \qquad v_y(0) = v_{y,o} \tag{5.8b}$$

and

$$\frac{dv_z}{dt} = 0\,. \quad v_z(0) = v_{z,o}\,. \tag{5.8c}$$

The interesting feature here is that $v_x(t)$ and $v_y(t)$ are mutually coupled but that $v_z(t)$ is independent of the former. Hence, the z-component is therefore straightforward and based on Eq.(5.8c), we have

$$v_z(t) = v_{z,o} = v_{\parallel} \tag{5.9}$$

which is a constant with the parallel line subscripts referring to the direction of the **B**-field. Then with the reference z-coordinate as $z(0) = z_o$, another integration gives

$$z(t) = z_o + v_{\parallel}t \tag{5.10}$$

with the trajectory of the charged particle motion in the z-direction thus established.

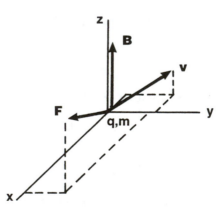

Fig. 5.3: Orientation of a magnetic field so that **B** is normal to the x-y plane and **F** is perpendicular to both **B** and **v**.

To solve for the velocity and position components perpendicular to the **B**-field direction–that is $v_x(t)$, $v_y(t)$, $x(t)$ and $y(t)$–we first uncouple Eqs.(5.8a) and (5.8b) by differentiation and substitution to obtain

$$\frac{d^2 v_x}{dt^2} = \left(\frac{q\,B_z}{m}\right)\frac{dv_y}{dt} = -\omega_g^2 v_x \tag{5.11a}$$

and

$$\frac{d^2 v_y}{dt^2} = -\left(\frac{q\,B_z}{m}\right)\frac{dv_x}{dt} = -\omega_g^2\, v_y \,, \qquad (5.11b)$$

where, upon recognizing that both Eq.(5.11a) and Eq.(5.11b) describe a harmonic oscillation with the same frequency which results in a circular motion of the particle given suitable initial conditions, we have introduced the gyration frequency–sometimes also called the gyrofrequency or cyclotron frequency–as

$$\omega_g = \frac{|q|\,B_z}{m}\,. \qquad (5.12)$$

Evidently, the following forms satisfy Eq.(5.11):

$$v_x(t) = A_x \cos(\omega_g t) + D_x \sin(\omega_g t)\,, \qquad (5.13a)$$

$$v_y(t) = A_y \cos(\omega_g t) + D_y \sin(\omega_g t) \qquad (5.13b)$$

with A_x, A_y, D_x and D_y constants to be determined for the case of interest.

It is known that a plasma possesses diamagnetic properties. This means that the direction of the magnetic field generated by the moving charged particles is opposite to that of the externally imposed field. Therefore, not only must the equations resulting from Eq.(5.6) be satisfied but the moving charged particle representing in fact a current flow along its trajectory must generate a magnetic field in opposition to **B**. Some thought and analytical experimentation with Eqs.(5.13) yields velocity components for a positive ion as

$$v_x(t) = v_o \cos(\omega_g t + \phi) \qquad (5.14a)$$

and

$$v_y(t) = -v_o \sin(\omega_g t + \phi) \qquad (5.14b)$$

where v_o is a positive constant speed and ϕ is the initial phase angle. For a negatively charged particle, Eq.(5.14b) will change its sign, while Eq.(5.14a) remains unchanged. Then, with $v_x(t)$ and $v_y(t)$ thus determined, a further integration yields x(t) and y(t) as

$$x(t) = x_o + \frac{v_o}{\omega_g}\sin(\omega_g t + \phi) \qquad (5.15a)$$

$$y(t) = y_o + sign(q)\frac{v_o}{\omega_g}\cos(\omega_g t + \phi) \qquad (5.15b)$$

describing thereby a circular orbit about the reference coordinate (x_o, y_o) for an arbitrary charge of magnitude q and sign designated by sign(q). This coordinate constitutes a so-called "guiding" centre which can, in view of Eq.(5.9), move at a constant speed in the z-direction. The gyration in 3-dimensions would therefore appear as a helix of constant pitch.

From Eqs.(5.14) it follows that the particle speed in the plane perpendicular to the magnetic field is a conserved quantity, and hence we write

$$v_o = \sqrt{v_x^2 + v_y^2} = v_\perp \,, \qquad (5.16)$$

in order to denote the velocity component perpendicular to **B**. Evidently, if $v_\parallel = 0$, the particle trajectory is purely circular, and the corresponding radius is obviously determined by

$$r_g = \frac{v_\perp}{\omega_g} = \frac{v_\perp m}{|q| B_z} .$$ (5.17)

The label gyroradius–as a companion expression to the gyrofrequency ω_g–is generally assigned to this term r_g although the name Larmor radius is also often used.

From the above it is evident that a heavy ion will have a larger radius than a lighter ion at the same $v_\perp$, and, by definition of ω_g, Eq.(5.12), an electron will possess a much higher angular frequency than a proton. Figure 5.4 depicts the circular trajectory of an ion and an electron as a projection of motion on to the x-y plane under the action of a uniform magnetic field $\mathbf{B} = B_z\mathbf{k}$. A useful rule to remember is that with $\mathbf{B}$ pointing into the plane of the page, an ion circles counterclockwise while electrons circle clockwise. Note that this is consistent with the diamagnetic property of a plasma by noting that, at the (x_o,y_o) coordinate, the self-generated $\mathbf{B}$-field points in the opposite direction to the external magnetic field vector $\mathbf{B}$.

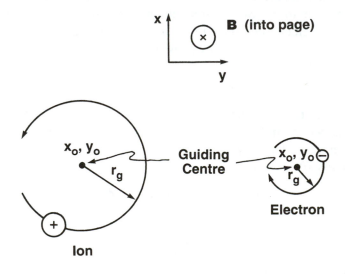

Fig. 5.4: Projection of ion and electron motion on the x-y plane which is perpendicular to the direction of the magnetic field vector.

With a constant velocity component parallel to the $\mathbf{B}$-field and a time-invariant circular projection in the perpendicular plane, the particle trajectory in 3-dimensions represents a helical pattern along the $\mathbf{B}$-line as suggested in Fig.5.5; note that $v_{z,o} = v_\parallel$ solely determines the axial motion. The pitch of the helix is also suggested in Fig.5.5 and is given by

$$\Delta z = z\left(t + \frac{2\pi\, r_g}{v_\perp}\right) - z(t)\,. \tag{5.18}$$

Using Eq.(5.10) gives

$$\Delta z = \left[z_o + v_\parallel\left(t + \frac{2\pi\, r_g}{v_\perp}\right)\right] - \left[z_o + v_\parallel t\right]$$

$$= 2\pi\left(\frac{v_\parallel}{v_\perp}\right) r_g \tag{5.19}$$

$$= 2\pi\frac{v_\parallel}{\omega_g}$$

which is constant for our case. Thus, an isolated charged particle in a constant, homogeneous magnetic field traces out a helical trajectory of constant radius, frequency and pitch.

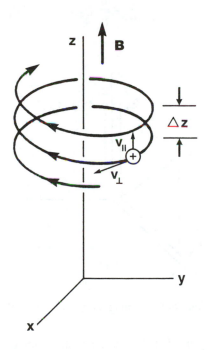

Fig. 5.5: Trajectory of an isolated ion in a uniform magnetic field.

5.4 Combined Electric and Magnetic Field

Consider next the effect of a combined **E** and **B** field on the trajectory of a charged particle. With both fields uniform and constant throughout our space of interest, the trajectory is expected to be more complex since it will combine features of the two trajectories of Figs. 5.2 and 5.5.

With the **B**-field and **E**-field possessing arbitrary directions, it is useful to orient the Cartesian coordinate systems so that the z-axis is parallel to **B** with **E** in the x-z plane, Fig. 5.6. The equation of motion, Eq.(5.1), is now expanded to give

$$m\frac{d}{dt}\left(v_x\mathbf{i}+v_y\mathbf{j}+v_z\mathbf{k}\right)=q\left[\left(E_x\mathbf{i}+E_z\mathbf{k}\right)+\left(v_x\mathbf{i}+v_y\mathbf{j}+v_z\mathbf{k}\right)\times\left(B_z\mathbf{k}\right)\right] \quad (5.20)$$

so that the velocity components must satisfy

$$\frac{dv_x}{dt}=\frac{q}{m}E_x+\frac{q}{m}B_z v_y=\frac{q}{m}E_x+\omega_g v_y\,sign(q)\,, \quad (5.21a)$$

$$\frac{dv}{dt}=-\frac{q}{m}B_z v_x=-\omega_g v_x\,sign(q), \quad (5.21b)$$

and

$$\frac{dv_z}{dt}=\frac{q}{m}E_z\,. \quad (5.21c)$$

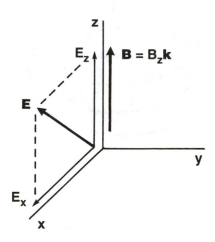

Fig. 5.6: Orientation of a magnetic field **B** and electric field **E**.

Here, the gyrofrequency ω_g, Eq.(5.12), has again appeared and the charge sign of the particle is indicated by sign (q), thereby specifying the direction of gyration. The velocity components v_x and v_y may be uncoupled by differentiation and substitution of Eqs.(5.21a) and (5.21b):

$$\frac{d^2 v_x}{dt^2}=0+\omega_g\frac{dv_y}{dt}sign(q)=-\omega_g^2 v_x \quad (5.22a)$$

and

$$\frac{d^2 v_y}{dt^2} = -\omega_g \frac{dv_x}{dt} sign(q)$$

$$= -\omega_g sign(q)\left(\frac{q}{m} E_x + \omega_g v_y sign(q)\right) \tag{5.22b}$$

$$= -\omega_g^2\left(\frac{E_x}{B_z} + v_y\right).$$

With E_x, B_z, and ω_g as constants, this last expression may therefore be written as

$$\frac{d^2 v_y^*}{dt^2} = -\omega_g^2 v_y^* \tag{5.23}$$

where

$$v_y^* = \frac{E_x}{B_z} + v_y . \tag{5.24}$$

The useful result here is that with this definition of v_y^*, the problem of determining the particle trajectory in a combined **E** and **B**-field, Eqs.(5.22a) and (5.23), has been cast into the mathematical framework of a **B**-field acting alone as in the preceding section. Using Eqs.(5.14) and (5.16), we find that

$$v_x = v_o^* \cos(\omega_g t + \phi^*) \tag{5.25a}$$

and

$$v_y^* = -v_o^* \sin(\omega_g t + \phi^*) sign(q) \tag{5.25b}$$

where the speed is now

$$v_o^* = \sqrt{v_{x,o}^2 + (v_{y,o} + E_x/B_z)^2} \tag{5.26}$$

and ϕ^* relates to the previous initial phase angle by

$$v_o^* \cos \phi^* = \sqrt{v_{x,o}^2 + v_{y,o}^2} \cos \phi . \tag{5.27}$$

Substituting for v_y^* in Eq. (5.24) and further introducing the solution for v_z, as evident from Eq.(5.21c), yields the following set of velocity components:

$$v_x = v_o^* \cos(\omega_g t + \phi^*), \tag{5.28a}$$

$$v_y = -v_o^* \sin(\omega_g t + \phi^*) sign(q) - \frac{E_x}{B_z} \tag{5.28b}$$

and

$$v_z = v_{\parallel} + \left(\frac{q E_z}{m}\right) t . \tag{5.28c}$$

The associated (x,y,z) coordinates follow from a further integration to yield

$$x = x_o + \frac{v_o^*}{\omega_g} \sin(\omega_g t + \phi^*), \tag{5.29a}$$

$$y = y_o + \frac{\overset{*}{v_o}}{\omega_g} \cos(\omega_g t + \phi^*) sign(q) - \frac{E_x}{B_z} t \qquad (5.29b)$$

and

$$z = z_o + v_\| t + \frac{1}{2}\left(\frac{q E_z}{m}\right) t^2 . \qquad (5.29c)$$

These equations, Eqs.(5.28) and (5.29), show that the charged particle still retains its cyclic motion but that it drifts in the negative y-direction for $E_x > 0$, Fig. 5.7, and oppositely if $E_x < 0$. Hence, the guiding centre's drift is in the direction perpendicular to both the **B**-field and **E**-field and in contrast to the preceding ($\mathbf{E} = 0, \mathbf{B} \neq 0$) case, ion and electron motion across the field lines now occur. Additionally, the pitch increases with time–Eqs.(5.28c) and (5.29c)–so that the trajectory resembles a slanted helix with increasing pitch.

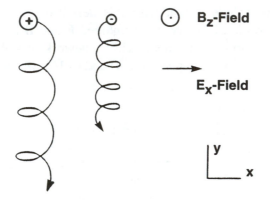

Fig. 5.7: Positive ion and electron drift in a combined uniform magnetic and electric field.

We now consider a generalization of the drift velocity caused by an arbitrary force **F** on a charged particle moving in an uniform **B**-field. To begin, we decompose the particle velocity into the components

$$\mathbf{v} = \mathbf{v}_{gc} + \mathbf{v}_g \qquad (5.30)$$

where $\mathbf{v}_{gc}$ is the velocity of guiding centre motion and $\mathbf{v}_g$ is the velocity of gyration relative to the guiding centre. The equation of motion is now written as

$$m\frac{d\mathbf{v}_{gc}}{dt} + m\frac{d\mathbf{v}_g}{dt} = \mathbf{F} + q(\mathbf{v}_{gc} \times \mathbf{B}) + q(\mathbf{v}_g \times \mathbf{B}) . \qquad (5.31)$$

Evidently the terms describing the circular motion, i.e. the second term on the left and the third on the right cancel one another according to Eq.(5.6). Further, in a static field exhibiting straight **B**-field lines the charged particle motion will be such that, averaged over one gyroperiod τ, the total acceleration perpendicular to **B** must vanish; that is, the guiding centre is not accelerated in any direction perpendicular to **B**. Therefore, we average the transverse part of Eq.(5.31) over this gyroperiod $\tau = 2\pi/\omega_g$ to write

$$0 = \frac{1}{\tau}\int_0^\tau \left(m \frac{d\,\mathbf{v}_{gc}}{dt} \right)_\perp dt = \frac{1}{\tau}\int_0^\tau \left(\mathbf{F}_\perp + q(\,\mathbf{v}_{gc} \times \mathbf{B}\,) \right) dt \, . \tag{5.32}$$

This now specifies

$$\overline{\mathbf{F}}_\perp = -q\left(\overline{\mathbf{v}}_{gc} \times \mathbf{B} \right) \tag{5.33}$$

with $\overline{\mathbf{F}}_\perp$ representing the gyro-period averaged force acting perpendicular to $\mathbf{B}$ and $\overline{\mathbf{v}}_{gc}$ the average velocity of the guiding centre, respectively. Taking next the cross-products with $\mathbf{B}$, we obtain

$$\overline{\mathbf{F}} \times \mathbf{B} = \overline{\mathbf{F}}_\perp \times \mathbf{B} = q\left[\overline{\mathbf{v}}_{gc}\, B^2 - \mathbf{B}\left(\overline{\mathbf{v}}_{gc} \cdot \mathbf{B} \right) \right] \tag{5.34}$$

where a vector product identity has been utilized for the last step. Note from Eq.(5.31) that the motion of the guiding centre parallel to the magnetic field $\mathbf{B}$ can only be affected by the parallel component of $\mathbf{F}$. Then, since $\mathbf{v}_g$ is perpendicular to $\mathbf{B}$, the velocity component $v_{gc,\parallel} = v_\parallel$ and is readily seen to be given by

$$v_{gc,\parallel}(t) = v_{gc,\parallel}(0) + \frac{1}{m}\int F_\parallel dt \, . \tag{5.35}$$

Of greater interest here is the perpendicular component $v_{gc,\perp}$ which is found from Eq.(5.34) as

$$\overline{\mathbf{v}}_{gc,\perp} = \frac{\overline{\mathbf{F}} \times \mathbf{B}}{q\,B^2} \tag{5.36}$$

because $(\,\overline{\mathbf{v}}_{gc,\perp} \cdot \mathbf{B}\,) = 0$. This identifies what is called the drift velocity due to the force $\mathbf{F}$, i.e.

$$\overline{\mathbf{v}}_{gc,\perp} = \mathbf{V}_{DF} \, . \tag{5.37}$$

This generalization allows us to specify an expression if the additional force acting in a static magnetic field domain is an electric force $\mathbf{F} = q\mathbf{E}$, so that the drift velocity due to a stationary electric field is

$$\mathbf{V}_{D,E} = \frac{\mathbf{E} \times \mathbf{B}}{B^2} \, . \tag{5.38}$$

The label "$\mathbf{E} \times \mathbf{B}$ drift" is frequently used in this context. Note that the drift is evidently independent of the mass and charge of the particle of interest, and consistent with the drift velocity term in Eq.(5.28b), i.e. $(- E_x / B_z)$.

5.5   Spatially Varying Magnetic Field

Each of the preceding cases involved a $\mathbf{B}$-field taken to be uniform throughout the space of interest. Such an idealization is, in practice, impossible although it can be approached to some degree by using sufficiently large magnet or coil arrangements. In general, however, it often becomes necessary to incorporate a space dependent **B**-field which governs the particles' motion. The associated equation of motion is hence written as

$$m\frac{d\mathbf{v}}{dt} = q[\mathbf{v} \times \mathbf{B}(\mathbf{r})].$$ (5.39)

A comprehensive analysis of this equation for a most general case may obscure some important instructional aspects so that a more intuitive approach has merit; this approach also provides for a convenient association with the preceding analysis while still leading to some useful generalizations.

We consider therefore a dominant magnetic field in the z-direction characterized by an increasing strength in the y-direction; that is, a constant non-zero gradient of the magnetic field strength, ∇B, exists everywhere given by

$$\nabla B = \frac{\partial B_z}{\partial y}\,\mathbf{j}.$$ (5.40)

Figure 5.8 provides a graphical depiction of this **B**-field showing also a test ion with specified initial velocity components.

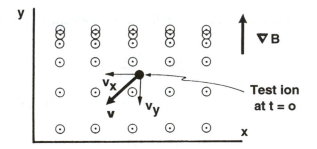

Fig. 5.8: Illustration of a spatially varying magnetic field **B** with the field lines pointing vertically up from the page.

From the previous analysis of the motion of a charged particle in a uniform **B**-field, it is known that the radius of gyration in the plane perpendicular to the local **B**-field is given by

$$r_g = \frac{v_\perp}{\omega_g} = \frac{v_\perp m}{|q|B}$$ (5.41)

where B is the local magnetic field magnitude at the point where the charged particle's guiding centre is located; thus, the gyroradius r_g varies inversely with B. Any motion in the z-direction will be additive and depends upon the initial $v_\parallel$ component; for reasons of algebraic simplification and pictorial clarity, we may therefore take $v_\parallel = 0$.

Referring to Eq.(5.41) and Fig. 5.8, it is evident therefore that an isolated charged test particle will trace out a trajectory whose radius of curvature becomes larger in the weaker magnetic field and smaller where the **B**-field is stronger. We can translate this effect into a process which tends to transform a circular path into a shifting cycloid-like path with the consequence that the particle trajectory now tends to cross magnetic field lines as suggested in Fig. 5.9.

Note that the radius of gyration, Eq.(5.41), is directly proportional to the mass of the particle and that the sign of the charge enters into the direction of the trajectory. This is also indicated in Fig. 5.9 for a positively and negatively charged particle, although not to scale. The feature that the gradient causes a drift of the positive ions in one direction and for electrons in the opposite direction has important consequences because the resulting charge separation introduces an electric field which, in a plasma, can contribute to plasma oscillations and instability, which we will further discuss in subsequent chapters.

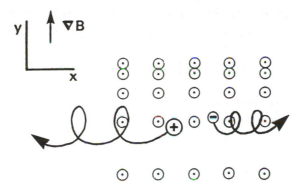

Fig. 5.9: Motion of a positively and negatively charged particle in a spatially varying magnetic field.

With an intuitive approach of the gradient-B drift phenomenon thus suggested, we consider next an approximate though useful analytical representation. The first approximation to be introduced is that for the case of a weak magnetic field inhomogeneity over distances corresponding to a gyroradius, that is, $|\nabla B|/B \ll 1/r_g$, only the first two terms of a Taylor expansion of $B_z(y)$ are retained, i.e.

$$B_z(y) \approx B_z(y_o) + (y - y_o)\frac{dB_z}{dy}\bigg|_{y=y_o} \tag{5.42}$$

where y_o corresponds to the y-coordinate of the guiding centre. Extending to the vector form for Eq.(5.42), we write

$$\mathbf{B}(y) \approx \mathbf{B}_o + (y - y_o)|\nabla B|\frac{\mathbf{B}_o}{B_o} \qquad (5.43)$$

where now $\mathbf{B}_o = \mathbf{B}(y_o)$. We next expand the equation of motion, Eq.(5.39), and write

$$m\frac{d\mathbf{v}}{dt} = q(\mathbf{v} \times \mathbf{B}_o)\left[1 + (y - y_o)\frac{|\nabla B|}{B_o}\right]. \qquad (5.44)$$

Recalling our preceding discussion of a drift caused by a general force $\mathbf{F}$ applied to a charged particle in a uniform B-field, we decompose $\mathbf{v}$ as in Eqs.(5.30) and (5.31) to write

$$m\frac{d\mathbf{v}_{gc}}{dt} = q(\mathbf{v}_{gc} \times \mathbf{B}_o) + q\frac{|\nabla B|}{B_o}(y - y_o)\left[(\mathbf{v}_{gc} + \mathbf{v}_g) \times \mathbf{B}_o\right] \qquad (5.45)$$

with the terms describing gyromotion in a uniform field $\mathbf{B}_o$ having been canceled. Comparison of Eq.(5.45) with Eq.(5.31) makes it evident that the last term in Eq.(5.45) can be identified as the force term $\mathbf{F}$ responsible for the ∇B drift. Following our analysis involving Eqs.(5.32) to (5.36), we need $\mathbf{F}$–which points in a direction perpendicular to $\mathbf{B}_o$–to be averaged over a gyroperiod τ so that

$$\overline{\mathbf{F}} = q\frac{|\nabla B|}{B_o} \cdot \frac{1}{\tau}\int_0^\tau [y(t) - y_o]\,[\mathbf{v}(t) \times \mathbf{B}_o]dt\,. \qquad (5.46)$$

Further, as a first order approximation of the product $|\nabla B|(y(t) - y_o)\mathbf{v}(t)$ in the above expression, we use for $y(t)$ and $v(t)$ the (undisturbed) uniform B-field solution of Eqs.(5.9), (5.14) and (5.15b) to write

$$y(t) - y_o \approx sign(q)\frac{v_\perp}{\omega_g}\cos(\omega_g + \phi) \qquad (5.47)$$

and

$$\mathbf{v}(t) \approx \begin{pmatrix} v_\perp \cos(\omega_g t + \phi) \\ -sign(q)v_\perp \sin(\omega_g t + \phi) \\ v_\parallel \end{pmatrix}. \qquad (5.48)$$

Upon substitution of this expression into Eq.(5.46) and performing the vector cross product–recalling also that $\mathbf{B}_o$ points into the positive z-direction–we obtain

$$\overline{\mathbf{F}} = q\frac{|\nabla B|}{B_o}\frac{1}{\tau}\int_0^\tau \frac{v_\perp^2}{\omega_g}B_o\cos(\omega_g t + \phi)\begin{pmatrix} -\sin(\omega_g t + \phi) \\ -sign(q)\cos(\omega_g t + \phi) \\ 0 \end{pmatrix}dt$$

$$\qquad (5.49)$$

$$= q\frac{|\nabla B|}{B_o}\left[-sign(q)\frac{\omega_g}{2\pi}\int_0^{2\pi/\omega_g}\frac{v_\perp^2}{\omega_g}B_o\cos^2(\omega_g t + \phi)dt\right]\mathbf{j}$$

where $\tau = 2\pi/\omega_g$ has been substituted and $\mathbf{j}$ is the unit vector in the y-direction;

note that integration of the product sin( )cos( ) provides no contribution by reason of orthogonality. Hence, integration of Eq. (5.49) leads to

$$\overline{\mathbf{F}} \approx -|q| \left(\frac{dB}{dy} \right) \left(\frac{v_\perp^2}{2\,\omega_g} \right) \mathbf{j} = -|q| \frac{v_\perp^2}{2\,\omega_g} \nabla B . \tag{5.50}$$

This latter expression is a general formulation since the y-axis was arbitrarily chosen due to the geometry of Sec. 5.3.

Finally, recalling Eqs.(5.36) and (5.37), we find the grad-B drift velocity as

$$\mathbf{v}_{D,\nabla B} = sign(q) \frac{v_\perp^2}{2\,\omega_g} \frac{\mathbf{B} \times \nabla B}{B^2} \tag{5.51}$$

and thus the positive ions and negative electrons drift across B-field lines in opposite directions in a non-uniform magnetic field as displayed in Fig. 5.9.

In subsequent chapters, we will discover that fusion devices which utilize magnetic fields for confinement of the reacting ions involve very complex magnetic field topologies. Not only are there electric fields and spatially varying magnetic fields present–thus giving rise to the drift velocities $\mathbf{v}_{D,E}$ and $\mathbf{v}_{D,\nabla B}$ just discussed–but there are additional complexities due to curvature in the magnetic field lines and gradients along these field lines. We thus now formulate expressions for the effective forces and resulting drift velocities due to these inhomogeneities in the B-fields, analogous to Eqs.(5.38) and (5.51).

5.6 Curvature Drift

Consider then charged particle motion in a magnetic field **B** whose field lines possess a radius of curvature R. In examining such a case we specify here the curved magnetic field lines by

$$\mathbf{B} = B_o \mathbf{k} + B_o \frac{z}{R} \mathbf{j} \tag{5.52}$$

when referred to in the vicinity of the y-axis, that is for |z/R| << 1. A graphical depiction is provided in Fig. 5.10.

Introducing this field configuration, Eq. (5.52), into the Lorentz-force equation, Eq.(5.6), yields

$$m \frac{d\mathbf{v}}{dt} = q(\mathbf{v} \times B_o \mathbf{k}) + q \left(\mathbf{v} \times \frac{z}{R} B_o \mathbf{j} \right) \tag{5.53}$$

which–upon decomposition according to Eq. (5.30) and subsequent cancellation of the gyration terms–reads as

$$m \frac{d\,\mathbf{v}_{gc}}{dt} = q\,B_o \left(\mathbf{v}_{gc} \times \mathbf{k} \right) + q\,B_o \frac{z}{R} (\mathbf{v} \times \mathbf{j}) . \tag{5.54}$$

For small curvatures, i.e. |z/R| << 1, we may, as in the previous cases, approximate the product $z(\mathbf{v} \times \mathbf{j})$ by the undisturbed uniform solutions given in

Eqs.(5.10) and (5.48), respectively, and thus obtain

$$\frac{d\,\mathbf{v}_{gc}}{dt} = \frac{q\,B_o}{m}\left(\mathbf{v}_{gc}\times\mathbf{k}\right) + \frac{q\,B_o\,v_{\|}t}{mR}\left[-v_{\|}\mathbf{i} + v_{\perp}\cos(\omega_g t)\mathbf{k}\right] \tag{5.55}$$

where we specified $z_o = 0$ and the initial phase angle of Eq. (5.48) as $\phi = 0$. Further substituting qB_o/m by Eq. (5.12), we separate Eq. (5.55) into its Cartesian components and write

$$\frac{dv_{gc,x}}{dt} = sign(q)\ \omega_g\left(v_{gc,y} - \frac{v_{\|}^2}{R}t\right) \tag{5.56a}$$

$$\frac{dv_{gc,y}}{dt} = -sign(q)\,\omega_g\,v_{gc,x} \tag{5.56b}$$

$$\frac{dv_{gc,z}}{dt} = sign(q)\,\omega_g\,\frac{v_{\|}v_{\perp}}{R}t\cos(\omega_g t)\,. \tag{5.56c}$$

Apparently, Eq. (5.56c) can be solved straightforwardly upon knowing that $v_{gc,z}(0) = v_{\|}$. Averaging the resulting velocity $v_{gc,z}$ over a gyration period will suppress the small oscillations with the cyclotron frequency ω_g and indicate a steady drift anti-parallel to the **B**-field; this velocity drift, however, is seen to depend on the initial condition, $v_{\perp} = v_o$, and is therefore not classified as a drift.

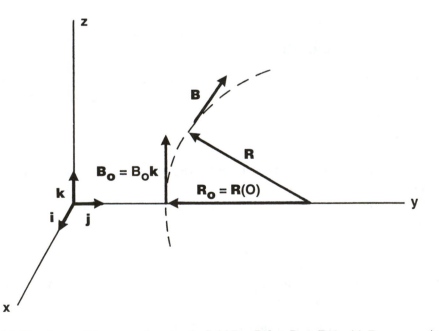

Fig. 5.10: Topology of the curved magnetic field $\mathbf{B} = B_o\mathbf{k} + B_o(z/R)\mathbf{j}$ with **R** representing the radius of curvature vector.

To solve for the velocity components perpendicular to **B** we uncouple the mutual dependencies of Eqs. (5.56a) and (5.56b) by differentiation and subsequent substitution to obtain

$$\frac{d^2 v_{gc,x}}{dt^2} + \omega_g^2 \, v_{gc,x} = -sign(q) \omega_g \frac{v_\parallel^2}{R} \, . \tag{5.57}$$

The solution of this inhomogeneous differential equation is accomplished by summing of the homogeneous solution of the form $A\cos(\omega_g t + \alpha)$ and a stationary solution to the inhomogeneous equation, which, by inspection of Eq.(5.57), we take as $(sign(q)v_\perp^2/\omega_g R)$. Referring to the initial conditions $v_{gc,x}(0)=v_{gc,y}(0)=0$ associated with the undisturbed field at $z(0) = 0$, we determine the constants A and α and finally find

$$v_{gc,x} = sign(q)\frac{v_\parallel^2}{\omega_g R}[\cos(\omega_g t) - 1] \tag{5.58a}$$

and

$$v_{gc,y} = \frac{v_\parallel^2}{\omega_g R}[\omega_g t - \sin(\omega_g t)] \, . \tag{5.58b}$$

The resulting drifts are, as preceedingly, revealed by averaging over a gyration period and are thus

$$\overline{\mathbf{v}_{gc,x}} = -sign(q)\frac{v_\parallel^2}{\omega_g R}\,\mathbf{i} \tag{5.59a}$$

and

$$\overline{\mathbf{v}_{gc,y}} = \frac{v_\parallel^2}{R}\bar{t}\,\mathbf{j} = \mathbf{a}_{cp}\bar{t} \tag{5.59b}$$

where $\bar{t}$ represents an average length of time corresponding to one half of the interval of averaging. The latter equation clearly demonstrates that the guiding centre is accelerated in the direction of $\mathbf{j}$ at the magnitude $v_\parallel^2/R$ which, of course, is known to be the centripetal acceleration $\mathbf{a}_{cp}$ required to make the particle follow the curved $\mathbf{B}$-field line. The only real *steady* drift is then accounted for by (Eq. 5.59a), in which we substitute according to the geometry of Fig. 5.10

$$\mathbf{i} = \mathbf{j} \times \mathbf{k} = -\frac{\mathbf{R}_o}{R} \times \frac{\mathbf{B}_o}{B_o} \tag{5.60}$$

and reintroduce $\omega_g = |q|B_o/m$ to derive the curvature drift velocity as

$$\mathbf{v}_{D,R} = \overline{\mathbf{v}_{gc,x}} = \frac{mv_\parallel^2}{q\,B_o^2}\frac{\mathbf{R}_o \times \mathbf{B}_o}{R^2} \tag{5.61}$$

which is thus again dependent upon the sign of the charge. Fig. 5.11 illustrates this drift of an ion spiraling around a curved magnetic field line.

A curved $\mathbf{B}$-field is shown below to be accompanied by a gradient $\mathbf{B}$-field, thus both specific drift velocity's components add to yield the resultant drift. We remark that, although the expressions for the different drift velocities derived here are correct, the assumptions and approximations used in the derivation are inconsistent in the sense that the suggested nonuniform **B**-fields do not satisfy Maxwell's equations

$$\nabla \times \mathbf{B} = 0 \qquad (5.62)$$

and

$$\nabla \cdot \mathbf{B} = 0 . \qquad (5.63)$$

In order that these fundamental equations are met, it is seen that an inhomogeneous magnetic field has always to be associated with some curvature and vice versa. Hence grad-B drift and curvature drift will occur simultaneously.

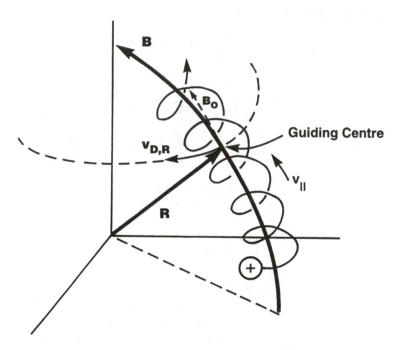

Fig. 5.11: Illustration of curvature drift of a positive ion moving along a curved **B**-field line.

5.7   Axial Field Variations

A magnetic field which varies spatially in strength in the direction of the field is of considerable importance in various mirror configurations–Fig. 4.2b. We suggest such a general case in Fig.5.12 and assess its effect on ion motion.

Cylindrical coordinates, (r,θ,z) in Fig. 5.12, will serve us well for which Maxwell's Second Equation, Eq.(5.63), gives

$$\nabla \cdot \mathbf{B} = \frac{1}{r} \frac{\partial}{\partial r} (r\, B_r) + \frac{1}{r} \frac{\partial B_\theta}{\partial \theta} + \frac{\partial B_z}{\partial z} = 0 . \qquad (5.64)$$

Axial symmetry is generally expected to hold so that $\partial B_\theta/\partial\theta = 0$ and we obtain

$$B_r = -\frac{1}{r}\int_0^r r'\,\frac{\partial B_z}{\partial z}dr' \approx -\frac{r}{2}\frac{\partial B_z}{\partial z} \qquad (5.65)$$

with the boundary condition B_r ($r = 0$) $= 0$ and $\partial B_z/\partial z$ taken to be independent of r. Then, assuming $B_\theta = 0$, we write

$$\mathbf{B} = B_r\,\mathbf{e}_r + B_z\,\mathbf{e}_z \qquad (5.66)$$

where $\mathbf{e}_r$ and $\mathbf{e}_z$ are the radial and axial unit vectors in cylindrical geometry, respectively, Fig. 5.12. The corresponding equation of motion of a charged particle in this field is

$$m\frac{d\mathbf{v}}{dt} = q(\mathbf{v}\times B_z\,\mathbf{e}_z) + q(\mathbf{v}\times B_r\,\mathbf{e}_r)\,. \qquad (5.67)$$

The first term is recognized as the force responsible for particle gyration about B_z and also as that which leads to a grad-B drift, e.g. Eq.(5.51), when the dependence of B_z on r becomes significant. The second term, explicitly written as

$$q(\mathbf{v}\times B_r\,\mathbf{e}_r) = -q\,v_\theta\,B_r\,\mathbf{e}_z + q\,v_z\,B_r\,\mathbf{e}_\theta \qquad (5.68)$$

contains a force parallel or antiparallel to the field on the axis and an azimuthal force. While the latter urges the particles to drift in the radial direction thus rendering their guiding centres to follow the specific magnetic field lines, the parallel force component will accelerate or, respectively, decelerate ions and electrons along the z-axis depending on the direction of their motion. With the aid of Eq. (5.65) this force component can be expressed as

$$F_\parallel = \tfrac{1}{2}q\,v_\theta\,r\,\frac{\partial B_z}{\partial z}\,. \qquad (5.69)$$

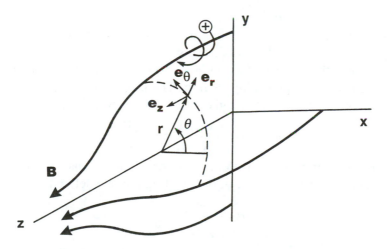

Fig. 5.12: Ion motion in a convergent magnetic field represented in cylindrical geometry.

We may now recognize the associated effect on the guiding centre by taking again an average $F_\parallel$ over a gyroperiod. Specifically taking the particles to gyrate around the central **B**-line at $r = 0$ renders the relation

$$v_\theta = -sign(q)v_\perp \tag{5.70}$$

so that for $r = r_g$, we obtain

$$F_\parallel = -\tfrac{1}{2}|q|v_\perp r_g \frac{\partial B_z}{\partial z} . \tag{5.71}$$

Upon substitution of r_g by Eq. (5.17), a generalization of this force can be shown to be given by

$$\mathbf{F}_\parallel = -\tfrac{1}{2}\frac{mv_\perp^2}{B}\nabla_\parallel B . \tag{5.72}$$

Thus a net force acts on charged particles–independent of their sign–in a direction opposite to that of increasing **B**.

5.8   Invariant of Motion

Constants of motion are important because they may reduce the number of independent variables in an equation characterizing some dynamic property of particles. In the absence of other force fields, the total energy E_o of a charged particle in a magnetic field is made up solely of the kinetic energy, since the only acting force, i.e. the Lorentz force, does not possess a potential energy; hence

$$E_o = \tfrac{1}{2}mv_\parallel^2 + \tfrac{1}{2}mv_\perp^2 \tag{5.73}$$

is a constant so that

$$\frac{d}{dt}\left(\tfrac{1}{2}mv_\parallel^2 + \tfrac{1}{2}mv_\perp^2\right) = 0 \tag{5.74}$$

where, as before, $v_\parallel$ and $v_\perp$ refer to velocity components parallel and perpendicular to the applied **B**-field direction.

Consider now the motion of a charged particle moving in a non-uniform **B**-field as described in the preceding section. Then, the parallel component of the force on this particle was seen to be given by the grad-B force, Eq. (5.72), as

$$\mathbf{F}_\parallel = m\frac{d\,v_\parallel}{dt} = -\frac{mv_\perp^2}{2B}\frac{\partial \mathbf{B}}{\partial s} = -\mu\frac{\partial \mathbf{B}}{\partial s} \tag{5.75}$$

where ds is a differential path element along **B** and μ has been introduced representing the magnetic moment of the gyrating particle defined by

$$\mu = \frac{\tfrac{1}{2}mv_\perp^2}{B} . \tag{5.76}$$

Further, since $v_\parallel = ds/dt$, we write,

$$\frac{\partial \mathbf{B}}{\partial s} = \frac{\partial \mathbf{B}}{\partial s}\cdot\frac{ds}{dt}\cdot\frac{1}{v_\parallel} = \frac{1}{v_\parallel}\frac{d\mathbf{B}}{dt} \tag{5.77}$$

with (dB/dt) describing the variations of **B** as seen by the particle moving with speed $v_\parallel$. Combining this relation with Eq.(5.75) provides

$$m\frac{dv_\parallel}{dt} = -\frac{\mu}{v_\parallel}\frac{dB}{dt} \qquad (5.78a)$$

or, respectively, in a rearranged form

$$\frac{d}{dt}\left(\tfrac{1}{2}mv_\parallel^2\right) = -\mu\frac{dB}{dt} . \qquad (5.78b)$$

We now insert this relationship into Eq.(5.74) together with the definition of μ, Eq.(5.76), to obtain

$$\left[-\mu\frac{dB}{dt}\right] + \left[\frac{d}{dt}(\mu B)\right] = 0 . \qquad (5.79)$$

Evidently, this relation holds only if

$$\frac{d}{dt}(\mu) = 0 \qquad (5.80)$$

as the individual isolated charged particle moves in the spatially varying magnetic field. That is, the magnetic moment does not vary with time.

Note that the grad-B force of Eq. (5.72), used here for establishing Eq. (5.80), had been derived by approximating the radial magnetic field component according to Eq. (5.65). Hence, for the case considered, the magnetic moment μ is conserved only in this approximation. From the definition of μ, Eq. (5.76), it is evident that μ is an invariant if **B** is constant. For magnetic fields slowly varying in space and/or in time, that is, relative changes in B are very small over a distance equal to the radius of gyration,

$$\frac{r_g}{B}\nabla_\parallel B \ll 1 , \qquad (5.81a)$$

and/or within a period of gyration,

$$\frac{1}{\omega_g B}\frac{dB}{dt} \ll 1 , \qquad (5.81b)$$

the magnetic moment is found to remain invariant in this first-order approximation. Consequently, μ is only approximately conserved and therefore called an adiabatic invariant.

The practical consequence of this is as it relates to the so-called magnetic mirror effect. As **B** increases toward the "throat" of the mirror region, $v_\perp$ must increase and, in view of Eq.(5.74), $v_\parallel$ must decrease; thus, ions tend to decelerate along the axial direction as they move into the higher magnetic field of the mirror throat where $v_\perp$ can even become zero, if B is high enough. Note that $F_\parallel$, which is given in Eq.(5.72) and which points opposite to ∇B, is still acting on the particles thus causing a reflection.

5.9   Cyclotron Radiation

We have established that the helical motion of a charged particle, guided by magnetic field lines as suggested in Fig. 5.5, involves a centripetal acceleration and therefore leads to the emission of radiation called cyclotron radiation and evidently involves an energy loss for the particle. We assess the associated power loss starting with the classical expression for the radiation emission rate of ions and electrons moving in an accelerating field which is known to exhibit the proportionality

$$P_{cyc} \propto N_i q_i^2 a_i^2 + N_e q_e^2 a_e^2 \approx N_e q_e^2 a_e^2 \tag{5.82}$$

where $a_{(\,)}$ denotes the respective accelerations and $q_{(\,)}$ the respective electrical charges with $q_i = |Zq_e|$. Here, the single particle radiation as given by Eq.(3.40) has been multiplied by the particle number density in order to provide an expression for power density. Note also that the smaller mass of the electrons will ensure that with their attendant higher acceleration–recall $a_e = F/m_e$–the electrons will be the predominant contributors to this cyclotron power loss. Ion cyclotron radiation is hence neglected. For cyclical motion in a magnetic field, the acceleration of electrons is a constant given by

$$a_e^2 = \left(\frac{v_{\perp e}^2}{r_{g,e}}\right)^2 = \left(\frac{v_{\perp e}^2}{r_{g,e}^2}\right) v_{\perp e}^2 \; . \tag{5.83}$$

Further, knowing the electron gyrofrequency as $\omega_{g,e} = |q_e|B/m_e$ with B as the controlling magnetic field, the relation of Eq. (5.41) allows us to write

$$\frac{v_{\perp e}^2}{r_{g,e}^2} = \frac{q_e^2 B^2}{m_e^2} \; . \tag{5.84}$$

Finally, assuming a Maxwellian distribution of electrons, we may take from kinetic energy considerations or Eq.(2.19b)

$$v_{\perp e}^2 \propto kT_e \tag{5.85}$$

giving therefore the cyclotron power density, Eq.(5.82) as,

$$P_{cyc} \propto N_e \left(\frac{q_e^4 B^2}{m_e^2}\right)(kT_e) \tag{5.86}$$

or

$$P_{cyc} = A_{cyc} N_e B^2 kT_e \tag{5.87}$$

where a constant of proportionality is explicitly introduced. With N_e in units of m^{-3}, B in units of Tesla, and kT in units of eV, the constant is given by $A_{cyc} \approx 6.3 \times 10^{-20}$ J·eV^{-1}· Tesla^{-2}·s^{-1} for P_{cyc} in units of W·m^{-3}. Thus, cyclotron radiation is most important at very high magnetic fields and high electron temperatures. Indeed, with electrons at a possibly very high energy, they may need to be subjected to a relativistic treatment.

Cyclotron radiation exists in the far infrared radiation spectrum with a wavelength of 10^{-3} - 10^{-4} m and is therefore partially re-absorbed in a plasma. Further, the emitted radiation may be reflected from the surrounding wall in a magnetic confinement fusion device and thereby re-enter the plasma. Hence, we choose to write the net cyclotron power finally lost from a plasma as

$$P_{cyc}^{net} = A_{cyc} N_e B^2 kT_e \psi \tag{5.88}$$

where ψ accounts for the complex processes of reflection and reabsorption of cyclotron radiation and is a dimensionless function involving several plasma parameters, including the magnetic field strength and the reflectivity of the surrounding wall. For a reasonable reflectivity of 90% and conditions expected in a fusion reactor, the range of ψ could typically be from 10^{-4} (for low T_e) to 10^{-2} (for very high T_e).

Problems

5.1   The maximum attainable magnetic field **B** is expected to be about 20 Tesla. For $v_\parallel \approx 0$, estimate the associated gyrofrequency, ω_g, and gyroradius, r_g, for electrons and protons given $kT_e = kT_i = 5$ keV.

5.2   Graphically depict the motion of an ion in a combined **E**-field and **B**-field. Display the results in an isometric representation.

5.3 (a) Determine the motion and position of a positive test charge in an electric field given by $\mathbf{E} = (E \cos(\omega t), 0, 0)$ with initial velocity $\mathbf{v}(0) = (v_{x,o}, 0, v_{z,o})$ and initial position at the origin.
    (b) For a similar positive test charge in only a magnetic field given by $\mathbf{B} = (B \cos(\omega t), 0, 0)$ with initial velocity $\mathbf{v}(0) = (v_{x,o}, v_{y,o}, v_{z,o})$ and initial position at the origin, describe qualitatively and sketch the motion for $0 < t < 2\pi/\omega$ and $\omega \ll \omega_g$.

5.4   Perform the integration suggested in Eq.(5.49).

5.5   Confirm the statement in Sec. 5.6 that integrating the solution of Eq. (5.56c), given appropriate initial conditions, over one gyration period yields a drift in the negative-$v_\parallel$ direction dependent upon $v_\perp$.

5.6   As discussed at the end of Sec. 5.6, curvature drift and ∇B-drift will always occur together. Using the co-ordinate system of Fig. 5.10, approximate–in analogy to Sec. 5.5 and 5.6–the magnetic field strength $\mathbf{B}(y,z) = (0, B_y(y,z), B_z(y,z))$ with a first order Taylor-expansion at the y-axis in order to account for both weak curvature and weak inhomogeneity (note that $\partial B_y/\partial y = \partial B_z/\partial z = 0$ at the reference point). What simple condition relating the degree of curvature an of inhomogeneity is then required to satisfy Maxwell's Eqs. (5.62) and (5.63)?

5.7   Demonstrate that the expression for the magnetic moment of an ion, Eq.(5.76), also follows from the basic definition of a current loop I encircling an area A (i.e., $\mu = IA$), and also identify the units for the magnetic moment μ.

5.8   Compare P_{cyc}^{net} with P_{br} for a deuterium-tritium plasma of density $N_e = N_i = 10^{20}$ m^{-3} confined by a magnetic field of magnitude $B = 6$ T.
Consider the specific cases:

   (i) $T_e = 10$ keV; $\psi = 10^{-3}$
   (ii) $T_e = 50$ keV; $\psi = 10^{-2}$
   and also compare $P_{br} + P_{cyc}^{net}$ with P_{fu} at the given conditions.

5.9   Consider a 50:50% MCF device having $T_i = T_e$ and a plasma beta value of 0.2, and for which $\psi \approx 10^{-3}$. Compute and compare the modified ignition temperature, i.e. the temperature at which $P_{br} + P_{cyc}^{net} = f_{c,dt} P_{dt}$, to T_{ign}^* of problem 3.6.

5.10   Calculate an expression for the power density of cyclotron radiation emitted from a d-h fusion plasma of density $N_i + N_e$ and of kinetic temperature $T_i \approx T_e = T$ by knowing that a particle of charge q and velocity $\mathbf{v}$ will–according to classical electromagnetic theory–emit radiation at a power $P_{rad} \propto q^2 |d\mathbf{v}/dt|^2$.
   (a) How much smaller is the cyclotron radiation power of ions than that of electrons?
   (b) Compare the primary cyclotron radiation power density (that with no reabsorption) to the fusion energy release per unit time for the case of $N_d = N_h = 2 \times 10^{20}$ m^{-3}, $T = 80$ keV and $B = 6$ Tesla, and discuss the importance of radiation reflecting walls in an MCF reactor.
   (c) Plot P_{cyc}^{net} (accounting for reflection and reabsorption by the approximation $\psi(T) \approx 10^{-4.2}$ [T(keV) / 1 keV]$^{1.4}$) and P_{br} as functions of T over the temperature range 10 to 120 keV.

6. Bulk Particle Transport

Media in which fusion reactions occur consist mainly of interspersed charged particles which are affected by short and long range forces. The cumulative effect of these forces combined with the intractability of an analytical description of each individual particle suggests that various approaches be used in the determination of the macroscopic behaviour of an ensemble of moving and colliding particles.

6.1   Particle Motion

The motion of a single particle of mass m is described by Newton's Law

$$m\frac{d\mathbf{v}}{dt} = \sum_j \mathbf{F}_j \tag{6.1}$$

where $\mathbf{F}_j$ is the j-th force vector acting on the particle and $\mathbf{v}$ is its velocity. For example, an isolated particle of mass m and charge q moving in a gravitational field $\mathbf{g}$, an electric field $\mathbf{E}$, and a magnetic field $\mathbf{B}$, has its space-time trajectory described by

$$m\frac{d\mathbf{v}}{dt} = m\mathbf{g} + q\mathbf{E} + q(\mathbf{v}\times\mathbf{B}). \tag{6.2}$$

While Eq.(6.2) is indeed very useful for some applications, it suffers from an overriding restriction: it describes the motion of an isolated particle only and thus excludes any possible interaction with other particles. This is indeed a severe restriction for fusion energy applications because power density requirements demand that about 10^{20} fusile ions be contained in one m^3; with such a large number of ions nearby each possessing a different velocity, it is evident that Coulomb interactions alone will lead to a most complex collection of time varying electrostatic forces acting on the particles. While in principle, one might specify a dynamical equation of the form of Eq.(6.1) for each particle, one would need perhaps 10^{20} force terms on the right hand side per unit volume; solving such equations for each of the interacting particles is, of course, totally unmanageable for computational purposes, due to the enormous number of these coupled equations and the lack of knowledge of individual initial conditions.

As an alternative to using an exceedingly large number of equations each containing many terms, it has been found that other approaches which are mathematically tractable provide, in selected applications, satisfactory agreement with experiments and a useful predictive quality. One frequently used conceptual approach is to consider a plasma as a multicomponent interpenetrating low-density fluid and then employ suitable continuum mechanics methods. Another approach is to use probabilistic considerations for each group of particle species and then perform an analysis based on methods of sampling statistical physics.

While no one approach is generally applicable to all cases of conceivable interest, a careful choice of conceptual constructs and methodologies can lead to descriptions which are remarkably useful in characterizing selected space-velocity-time aspects of a particular species' population in a fusion medium. It is important therefore to develop a good understanding of the imposed assumptions in order to recognize important restrictions of a particular conceptual development and its consequent mathematical description.

6.2   Continuity and Diffusion

As a first fluidic description of a medium containing fusion fuel ions and sustaining fusion reactions, we consider a characterization which emphasizes particle mobility. Consider therefore an arbitrary volume V containing several time-varying particle populations $N_1^*(t)$, $N_2^*(t)$, ..., $N_j^*(t)$, ..., each species characterized by some macroscopic kinetic property such as temperature. We take the volume's surface to be non-reentrant and of total area A with its outward normal direction determined by the differential vector d**A**.

Our interest here is in the $N_j^*(t)$ population which, in the case of a spatially varying number density $N_j(\mathbf{r},t)$, is found from

$$N_j^*(t) = \int_V N_j(\mathbf{r},t) d^3r \ . \tag{6.3}$$

In general, the population $N_j^*(t)$ can change with time because of various types of gain and loss reaction rates, $R_{\pm j}^*$, and because of inflow and outflow rates, $F_{\pm j}^*$, across the total surface A. Hence, the rate equation for $N_j^*(t)$ is evidently

$$\frac{dN_j^*}{dt} = \left(\sum_k R_{+jk}^* - \sum_l R_{-jl}^* \right) + \left(F_{+j}^* - F_{-j}^* \right), \tag{6.4}$$

with the subscripts k and l enumerating different reaction types.

To begin, we now restrict ourselves to the case for which, in any fractional volume ΔV of V, the reaction contributions are zero or that the reaction gains are exactly canceled by the reaction losses; that is, we take

$$\sum_k R_{+jk}^* - \sum_l R_{-jl}^* = 0 \ . \tag{6.5}$$

Additionally, we refer to the condition

$$F_{+j}^* = 0 \tag{6.6}$$

which means there is no fueling by injection or other mechanisms, and we determine the outflow rate across the surface, accounting for global particle leakage, via the local particle current vector $\mathbf{J}_j(\mathbf{r},t)$ through

$$F_{-j}^* = \int_A \mathbf{J}_j(\mathbf{r},t) \cdot d\mathbf{A} \ . \tag{6.7}$$

As previously introduced, $d\mathbf{A}$ is an oriented surface element pointing outward. Obviously, particles leak out where $\mathbf{J}|_A$ also points in this outward direction.

We now substitute Eqs.(6.3) and (6.5) - (6.7) into Eq.(6.4) to obtain for the space-time description of the j-type particle species

$$\frac{d}{dt} \int_V N_j(\mathbf{r},t) d^3r = -\int_A \mathbf{J}_j(\mathbf{r},t) \cdot d\mathbf{A} \tag{6.8}$$

for the case specified by Eqs. (6.5) and (6.6). Then, in order to reduce both integrals to the same integration variable, we use Gauss' Divergence Theorem

$$\int_A \mathbf{J}_j(\mathbf{r},t) \cdot d\mathbf{A} = \int_V \nabla \cdot \mathbf{J}_j(\mathbf{r},t) d^3r \tag{6.9}$$

so that Eq.(6.8), with the inclusion of differentiation under the integral sign, becomes

$$\int_V \left[\frac{\partial}{\partial t} N_j(\mathbf{r},t) + \nabla \cdot \mathbf{J}_j(\mathbf{r},t) \right] d^3r = 0 \ . \tag{6.10a}$$

Here, the partial derivative is used since the temporal change in $N_j(\mathbf{r},t)$ is considered at fixed spatial coordinates, respectively. Since the volume V is arbitrary, we must therefore have

$$\frac{\partial}{\partial t} N_j(\mathbf{r},t) + \nabla \cdot \mathbf{J}_j(\mathbf{r},t) = 0 \ . \tag{6.10b}$$

This important relation is known as the Continuity Equation and must hold everywhere in the volume of interest and for all times of relevance providing the assumptions imposed here hold. If, however, there were sources and sinks for the considered species j in the volume of interest–i.e. reactions which produce or consume j-type particles–or particle injection, the corresponding gain and loss rate densities would appear on the right hand side of Eq.(6.10b).

Though compact, Eq.(6.10b) however suffers from a severe shortcoming: it represents only one equation containing two unknown functions, the scalar particle density $N_j(\mathbf{r},t)$ and the vector particle current $\mathbf{J}_j(\mathbf{r},t)$. Evidently, another relationship between these two quantities is necessary. As it turns out, this is often possible.

Many particle fluid media possess the property that the local particle current $\mathbf{J}(\mathbf{r},t)$ is proportional to the negative particle density gradient, $-\nabla N(\mathbf{r},t)$. This is often called a diffusion phenomenon and associated with the label Fick's Law. Introducing a proportionality factor D, called the diffusion coefficient, we may write therefore everywhere in the medium

$$\mathbf{J}_j(\mathbf{r},t) = -D_j \nabla N_j(\mathbf{r},t) \tag{6.11}$$

so that, by substitution into Eq.(6.10b) we get

$$\frac{\partial}{\partial t} N_j(\mathbf{r},t) + \nabla \cdot \left[-D_j \nabla N_j(\mathbf{r},t) \right] = 0. \tag{6.12}$$

Numerous reductions are now often applicable. If the medium is homogeneous and isotropic then the diffusion coefficient is a space-independent scalar, and we obtain

$$\frac{\partial}{\partial t} N_j(\mathbf{r},t) - D_j \nabla^2 N_j(\mathbf{r},t) = 0. \tag{6.13}$$

Steady-state conditions in such a medium will yield Laplace's Equation

$$\nabla^2 N_j(\mathbf{r},t) = 0. \tag{6.14}$$

The diffusion coefficient in Eq.(6.12), D_j, is clearly of importance in specifying the spatial variation of the particle density $N_j(\mathbf{r},t)$. Considerable thought has gone into analytical characterizations of this parameter so that the diffusion processes be adequately incorporated. For example, for neutral particles in random thermal motion and in regions sufficiently far from boundaries, it has been found that, to a good approximation

$$D_j \propto \overline{v}_j \lambda_j \tag{6.15}$$

where $\overline{v}_j$ is the mean speed of the j-type particles and λ_j is the mean-free-path between scattering collisions among the j-type particles–and is much smaller than $(\nabla N/N)^{-1}$.

In contrast to neutral particles, charged particle motion in a magnetic field may involve considerable anisotropic diffusion governed by the local **B**-field requiring that a distinction between a direction which is parallel or perpendicular to the local **B**-field be made. Relationships with a good physical basis in classical diffusion are

$$D_{\parallel} \propto \left(\overline{v} \right)^2 \tau_c \tag{6.16a}$$

$$D_{\perp} \propto \frac{\left(\overline{v} \right)^2}{\omega_g^2 \tau_c}, \tag{6.16b}$$

where ω_g is the gyrofrequency and τ_c is the mean collision time.

By analogy to the mean-time between fusion events τ_{fu}, Eq.(4.3), we take herein

$$\tau_c \propto \frac{1}{\sigma_s v} \propto v^3 \tag{6.17}$$

and recall that the Coulomb cross section σ_s varies as v^{-4}, Eq.(3.15).

Further, substituting for the averaged square velocity by the thermal energy $E_{th} \propto kT$, or the kinetic temperature, respectively, and using the explicit expression for ω_g given in Eq.(5.12), we rewrite Eq.(6.16a) and (6.16b) to exhibit the proportionalities

$$D_{\parallel} \propto T^{5/2} \tag{6.18a}$$

and

$$D_{\perp} \propto \frac{1}{B^2 \sqrt{T}} . \tag{6.18b}$$

Thus, a higher temperature and an increasing magnetic field will reduce classical diffusion across the **B**-field lines. This conclusion must, however, be tempered by the recognition that collective oscillations in a plasma destroy some of the classical diffusion features. The plasmas of interest to nuclear fusion applications are severely affected by such collective processes through plasma instabilities which do not obey classical diffusion considerations.

The destabilizing oscillations induce turbulent phenomena which enhance diffusion across the magnetic field lines. Incorporation of these considerations leads to so-called Bohm diffusion characterized by

$$D_B \propto \frac{T}{B} \tag{6.19}$$

thereby placing greater demands on magnetic confinement at higher temperature.

Finally, we add that the Continuity Equation for the particle density $N_j(\mathbf{r},t)$, Eq.(6.13), translates into an equation for the mass density ρ_j by using $\rho_j(\mathbf{r},t) = N_j(\mathbf{r},t)m_j$,

$$\frac{\partial}{\partial t} \rho_j(\mathbf{r},t) - D_j \nabla^2 \rho_j(\mathbf{r},t) = 0 \tag{6.20}$$

and is generally found to be of considerable utility, particularly in conjunction with other relationships as we will show next.

6.3   Particle-Fluid Connection

The preceding analysis leads to a compact space-time description for the particle or mass density. Of particular interest to us now is a characterization of collective motion in a plasma. In such a description, the identity of individual particles is put aside and the plasma is characterized by the space-time changes of macroscopic variables such as bulk speed, temperature, and pressure defined in a fluidic context.

To begin with we take the Continuity Equation, Eq.(6.12), as a necessary equation for the particle density everywhere in space and time. That is, if another equation is developed in which $N_j(\mathbf{r},t)$ appears as an independent function, both equations must hold simultaneously and any solution methodology is expected to involve both equations.

To be specific, we consider a space-time ensemble of N_j particles each of mass m_j and charge q_j to which we assign an average velocity $\mathbf{V}_j$ over an infinitesimal volume containing N_j. A simplified one-dimensional analogy is the case of freeway traffic with cars moving at different speeds, all together amounting to some average speed of traffic motion which, nonetheless, may vary with time and location.

In order to provide some continuity to our preceding analysis, we suppose that an electric field $\mathbf{E}$ and magnetic field $\mathbf{B}$ act on this moving space-time ensemble of N_j particles, referred to as a fluid element. Its equation of motion is

$$N_j m_j \frac{d\mathbf{V}_j}{dt} = N_j q_j \mathbf{E} + N_j q_j (\mathbf{V}_j \times \mathbf{B}), \quad j = i,e. \tag{6.21}$$

Recall that N_j used here is determined by Eq.(6.12) everywhere in space and time.

Imposing our space-time variation on the ensemble-average-velocity we specify $\mathbf{V}_j$ as a function of space and time, that is

$$\mathbf{V}_j(\mathbf{r},t) = \mathbf{V}_j(x, y, z, t). \tag{6.22}$$

Performing the differentiation in Eq.(6.21), according to the chain rule gives specifically

$$\frac{d\mathbf{V}_j}{dt} = \frac{\partial \mathbf{V}_j}{\partial x}\frac{dx}{dt} + \frac{\partial \mathbf{V}_j}{\partial y}\frac{dy}{dt} + \frac{\partial \mathbf{V}_j}{\partial z}\frac{dz}{dt} + \frac{\partial \mathbf{V}_j}{\partial t} \tag{6.23a}$$

or, in more compact vector notation

$$\frac{d\mathbf{V}_j}{dt} = (\mathbf{V}_j \cdot \nabla)\mathbf{V}_j + \frac{\partial \mathbf{V}_j}{\partial t}. \tag{6.23b}$$

The two terms on the right hand side possess an easily established interpretation; clearly, the second term $\partial\mathbf{V}_j/\partial t$ is simply the acceleration of the N_j ensemble at a fixed coordinate point; then, the first term incorporates the process of spatial variation in the ensemble velocity much as in our one-dimensional analogy the traffic in one section of the freeway may move faster than that in another section; the term convection is often used for this kind of fluid motion. We insert Eq.(6.23b) into Eq.(6.21) and again use $\rho_j = N_j m_j$ to obtain a corresponding equation for the ensemble's motion

$$\rho_j \left[(\mathbf{V}_j \cdot \nabla)\mathbf{V}_j + \frac{\partial \mathbf{V}_j}{\partial t} \right] = \rho_j^c \mathbf{E} + \rho_j^c (\mathbf{V}_j \times \mathbf{B}) \tag{6.24}$$

where the mass density ρ_j satisfies the Continuity Equation, Eq.(6.20), and $\rho_j^c(\mathbf{r},t) = N_j(\mathbf{r},t)q_j$ is the respective charge density. Note that, because of electromagnetic interactions in a plasma, ρ_j^c, $\mathbf{E}$ and $\mathbf{B}$ are all related via Maxwell's Equation which, simultaneously, must also hold with Eq.(6.24).

At this juncture, we recognize another process which is generally important in a hot plasma: the thermal motion of particles makes them randomly enter and exit the fluid element considered, leading to local pressure gradients which act as an additional force in Eq.(6.24). While, in general, this force can be obtained from the divergence of a stress tensor, we consider here the case of an isotropic fluid, for which we add the term $-\nabla p_j$ to Eq.(6.24) in order to include this force effect; finally then, the motion of a fluid element is given by

$$\rho_j\left[\left(\mathbf{V}_j\cdot\nabla\right)\mathbf{V}_j+\frac{\partial\mathbf{V}_j}{\partial t}\right]=\rho_j^c\mathbf{E}+\rho_j^c\left(\mathbf{V}_j\times\mathbf{B}\right)-\nabla p_j \qquad (6.25)$$

wherein the local particle pressures are related to the respective mass density via the Equation of State

$$p_j=C\rho_j^{\gamma_j} \qquad (6.26)$$

where C is a constant and γ_j is the ratio of specific heats for constant pressure and constant volume ($\gamma_j=C_p/C_v$) referring to the j-th species.

An important point must be emphasized that concerns the electric and magnetic fields, **E** and **B**, when the above set of equations is evaluated for ions and electrons. As noted previously, it is required that Maxwell's Equations be satisfied for the medium of interest. The complete set of resultant equations for ρ_i, ρ_e, $\mathbf{V}_i$, $\mathbf{V}_e$, p_i, p_e, **E**, **B**–consisting of 16 simultaneous scalar equations– represents what is known as a self-consistent description of the plasma fluid approximation by the so called magnetohydrodynamic (MHD) equations. The imposition of initial and boundary conditions clearly leads to a nontrivial problem description.

6.4   Particle Kinetic Description

As a final characterization of a statistically varying population, we consider what is often called transport theory or a kinetic description. A compact derivation which leads to a defining equation for a particle density distribution function and which also includes collisional effects as well as source effects–that is, comprehensive transport of particles and their properties–is suggested in the following.

Consider an ensemble of N particles each characterized by a distinguishable velocity vector v_i and coordinate vector r_i at time t. For that we introduce a distribution function f_N represented by

$$f_N=f_N\left(\mathbf{r}_1,\mathbf{v}_1;\mathbf{r}_2,\mathbf{v}_2;\dots;\mathbf{r}_i,\mathbf{v}_i;\dots;\mathbf{r}_N,\mathbf{v}_N;t\right), \qquad (6.27)$$

which is able to specify the probability that at time t, the particle 1 is found about the coordinate r_1 with the velocity vector v_1 while the position and velocity of particle 2 are about r_2 and v_2, and the remaining (N-2) particles are similarly found at distinct coordinate intervals in the position-velocity space. With N as the number of particles of interest, each characterized by three Cartesian space coordinates and three Cartesian velocity components, this phase-space is evidently a 6N-dimensional phase-space.

A fundamental principle in the description of particle distributions is known as the Liouville Theorem; this theorem states that a volume element in the 6N-dimensional phase-space remains constant for the motion of particles obeying the Hamiltonian equation of motion. For the N-particle distribution function f_N, Eq.(6.27), this theorem leads to the Liouville Equation

$$\frac{df_N}{dt} = 0 \tag{6.28}$$

if particle creation and/or particle destruction does not occur. Performing the differentiation of Eq.(6.28), the explicit differentiation which f_N must satisfy is evidently

$$\left[\frac{\partial}{\partial t} + \sum_{i=1}^{N} \left(\mathbf{v}_i \cdot \frac{\partial}{\partial \mathbf{r}_i} + \frac{d\,\mathbf{v}_i}{dt} \frac{\partial}{\partial \mathbf{v}_i} \right) \right] f_N = 0 . \tag{6.29}$$

This differential equation is generally the starting point for an analysis of a many particle system. Note that the acceleration term $d\mathbf{v}_i/dt$ can be replaced by $\mathbf{F}_i/m_i$ via Newton's Law with $\mathbf{F}_i$ representing the total force acting on the i-th particle.

Equation (6.29) is now subject to integration over a suitable large set of coordinates $\mathbf{r}_i$ and velocities $\mathbf{v}_i$ in order to substantially reduce the dimensionality of the phase-space distribution function from f_N to, say f_r. Thus, for r << N, we obtain

$$f_r(\mathbf{r}_1, \mathbf{v}_1; \dots; \mathbf{r}_r, \mathbf{v}_r; t) = A_r \int f_N \prod_{i=r+1}^{N} d^3 r_i \, d^3 v_i \tag{6.30}$$

with A_r a normalization constant generally chosen by convention, e.g. $A_r = 1/(N - r)!$, found via permutational calculus.

We will find it useful to separate the total force $\mathbf{F}_i$ acting on the i-th particle into its internal (inter-particle) and external components, $\sum_{j=1}^{N} \mathbf{F}_{ij}$ and $\mathbf{F}_i^{ext}$. Further, assuming that f_N vanishes on the boundaries of the phase space and that the only velocity-dependent forces are of Lorentzian form as in Eq.(5.6), the integration of Eq.(6.29) over the 6(N - r) variables $\mathbf{r}_{r+1},\dots,\mathbf{r}_N,\mathbf{v}_{r+1},\dots,\mathbf{v}_N$ leads–in combination with Eq.(6.30) and using the normalization of A_r–to

$$\left[\frac{\partial}{\partial t} + \sum_{i=1}^{r} \left(\mathbf{v}_i \cdot \nabla_{ri} + \frac{\mathbf{F}_i^{ext} + \sum_{j=1}^{r} \mathbf{F}_{ij}}{m_i} \cdot \nabla_{vi} \right) \right] f_r = -\sum_{i=1}^{r} \frac{1}{m_i} \int \mathbf{F}_{i,r+1} \cdot \nabla_{vi} f_{r+1} d^3 r_{r+1} d^3 v_{r+1}$$

$$j \neq i , \quad r = 1, 2, \dots, N - 1 , \tag{6.31a}$$

where we replaced $\partial/\partial \mathbf{r}$ and $\partial/\partial \mathbf{v}$ by the gradient symbols

$$\nabla_r = \frac{\partial}{\partial x} \mathbf{i} + \frac{\partial}{\partial y} \mathbf{j} + \frac{\partial}{\partial z} \mathbf{k} \tag{6.31b}$$

and

$$\nabla_v = \frac{\partial}{\partial v_x}\mathbf{i} + \frac{\partial}{\partial v_y}\mathbf{j} + \frac{\partial}{\partial v_z}\mathbf{k} \;. \tag{6.31c}$$

Equation (6.31a) is known as the Bogolyubov-Born-Green-Kirkwood-Yvon hierarchy and closes only with the Liouville Equation, Eq.(6.28), since the equation for f_r contains f_{r+1}. To accommodate the cutting off of this sequence at a reasonably small r, a physical approximation for f_{r+1} is therefore needed in terms of $f_1,...,f_r$ in order to arrive at a solvable set of equations. From this set of r equations the r-1 equations may be used to eliminate $f_2,...,f_r$, leaving a single kinetic equation for the one-particle distribution function f_1.

We now consider the case r = 1 which is described by

$$\frac{\partial f_1}{\partial t} + \mathbf{v}_1 \cdot \nabla_{r_1} f_1 + \frac{\mathbf{F}_1^{ext}}{m_1} \cdot \nabla_{v_1} f_1 = -\frac{1}{m_1} \int \mathbf{F}_{12} \cdot \nabla_{v_1} f_2 \, d^3 r_2 \, d^3 v_2 \;. \tag{6.32}$$

Here, $f_1(\mathbf{r}_1,\mathbf{v}_1,t)$ is the one-particle distribution function used to determine the probability of finding particle 1 at time t about the position $\mathbf{r}_1$ with velocity $\mathbf{v}_1$; note however, that herein we have lost the corresponding information about the other particles. Further, Eq.(6.32) contains the two-particle distribution $f_2(\mathbf{r}_1,\mathbf{v}_1,\mathbf{r}_2,\mathbf{v}_2,t)$ which specifies the probability that at time t particle 1 moves with $\mathbf{v}_1$ about $\mathbf{r}_1$ and, simultaneously, particle 2 will be found about $\mathbf{r}_2$ with velocity $\mathbf{v}_2$, with no such individual information about the remaining N-2 particles.

Writing the one-particle distributions

$$f_1(\mathbf{r}_1,\mathbf{v}_1,t) = f(1) \tag{6.33a}$$

for particle 1,

$$f_1(\mathbf{r}_2,\mathbf{v}_2,t) = f(2) \tag{6.33b}$$

for particle 2, and expressing the two-particle distribution by cluster expansion as

$$f_2(\mathbf{r}_1,\mathbf{v}_1,\mathbf{r}_2,\mathbf{v}_2,t) = f(1)f(2) + C(1,2) \tag{6.33c}$$

with C representing the so-called pair correlation function relating kinetic variations of particles 1 and 2, most significantly within the close range of their interaction, we find upon substitution of these expansions the following kinetic equation:

$$\frac{\partial f(1)}{\partial t} + \mathbf{v}_1 \cdot \nabla_{r_1} f(1) + \frac{\mathbf{F}_1^{ext}}{m_1} \cdot \nabla_{v_1} f(1) + \frac{1}{m_1}\left(\int \mathbf{F}_{12} f(2) d^3 r_2 \, d^3 v_2 \right) \cdot \nabla_{v_1} f(1)$$

$$= -\frac{1}{m_1} \int \mathbf{F}_{12} \cdot \nabla_{v_1} C(1,2) d^3 r_2 \, d^3 v_2 \;. \tag{6.34}$$

The expression within brackets in the fourth term on the left-hand side represents the field force experienced by particle one due to the presence of the other particles and may be written as

$$\int \mathbf{F}_{12} f(2) d^3 r_2 \, d^3 v_2 = \mathbf{F}_1^{field} \;. \tag{6.35}$$

Actually, this is the particle contribution to the self-consistent field (for example:

an ion distribution generates an electromagnetic field which in turn maintains this distribution).

The right-hand side of Eq.(6.34) is the collision term for which some approximation must be applied to expand it also as a function of f(1) and f(2). Since various physical models exist for approximating the collision term, we choose to represent it here simply by $(\partial f(1)/\partial t)_c$. Thus the final kinetic equation for a one-particle distribution function is

$$\frac{\partial}{\partial t} f(\mathbf{r},\mathbf{v},t) + \mathbf{v} \cdot \nabla_r f + \frac{\mathbf{F}}{m} \cdot \nabla_v f = \left(\frac{\partial f}{\partial t}\right)_c \tag{6.36}$$

where $\mathbf{F}$ includes all external and field forces. Herein and subsequently we omit the subscript and $f(\mathbf{r},\mathbf{v},t)$ will refer to the one-particle distribution function which is representative of any one of the considered N particles of the same species, however does not provide information which distinguishes the individual kinetic behaviour of the particles. For charged particles moving in an electric and magnetic field, the total force is evidently

$$\mathbf{F} = q(\mathbf{E} + \mathbf{v} \times \mathbf{B}) \tag{6.37}$$

incorporating self-consistent fields.

If particle production and/or destruction also take place and particles can leave and/or enter the reaction volume under consideration, then we account for these sources and sink processes on the right-hand side of Eq.(6.36) by an additional term $(\partial f/\partial t)_s$ to be specified according to the appropriate conditions.

In the statistical sense $[f(\mathbf{r},\mathbf{v},t)\, d^3r\, d^3v]$ is interpreted as the probability of finding at time t a particle of the species of interest in the differential volume element $d^3r\, d^3v$ about the point $(\mathbf{r},\mathbf{v})$ in the 6-dimensional phase space, Fig.6.1. We may now conclude that the expected number of such particles in this differential volume element is

$$N(\mathbf{r},\mathbf{v},t)d^3r\,d^3v = N^* f(\mathbf{r},\mathbf{v},t)d^3r\,d^3v \tag{6.38}$$

where N^* is the total number of particles in the entire volume at some reference time. Hence

$$N(\mathbf{r},\mathbf{v},t)d^3r\,d^3v = \left(\begin{array}{l}\text{expected number of particles in } d\mathbf{r} \\ \text{about } \mathbf{r} \text{ and in } d\mathbf{v} \text{ about } \mathbf{v} \text{ at time t}\end{array}\right). \tag{6.39}$$

Obviously the normalization applied to Eq.(6.39) is

$$\iint f(\mathbf{r},\mathbf{v},t)d^3r\,d^3v = \frac{N(t)}{N^*} \tag{6.40}$$

with the number density obtained by

$$N(\mathbf{r},t) = \int N(\mathbf{r},\mathbf{v},t)d^3v = N^* \int f(\mathbf{r},\mathbf{v},t)d^3v. \tag{6.41}$$

The transport equation for a charged particle density distribution $N(\mathbf{r},\mathbf{v},t)$ in an electric and magnetic field and subjected to collisional effects as well as to particle production/destruction is therefore finally written as

$$\frac{\partial N}{\partial t} + \mathbf{v} \cdot \nabla_r N + \frac{q}{m}(\mathbf{E} + \mathbf{v} \times \mathbf{B}) \cdot \nabla_v N = \left(\frac{\partial N}{\partial t}\right)_c + \left(\frac{\partial N}{\partial t}\right)_s . \qquad (6.42)$$

A solution analysis of this equation for N = N(**r**,**v**,t) requires the self-consistent specification of **E** and **B** (determined in conjunction with Maxwell's Equations) everywhere, identification of the form of the collision and source/sink terms and the imposition of initial and boundary conditions. A reaction domain will generally contain more than one species of particles so that the determination of the overall collision kinetics requires several equations of the form of Eq.(6.42)– one for each species. All of these may be coupled by several collision terms since collisions will invariably occur among the same and other species. Needless to say, the dimensionality of this problem is formidable. A simplification is possible by a reduction of the number of independent variables and elimination of some terms depending upon the problem of interest.

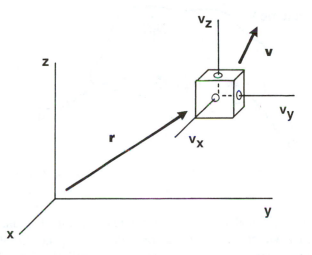

Fig. 6.1: Depiction of a 6-dimensional coordinate-velocity phase space volume element.

It is useful to note that the left-hand side of Eq.(6.42) constitutes, by our classification of processes, continuous changes in phase space while the right-hand parts are discontinuous changes. Thus, particle collisions, particle reactions and particle destructions are viewed as discontinuous–that is discrete processes– whereas particle acceleration can vary smoothly in the coordinate-velocity space. If only binary collisions are considered, the analysis is labeled "classical" or "neoclassical" if applied to toroidal geometry. As plasma wave-particle interactions become dominant such that the transport of mass, momentum or energy results therefrom, then the label "anomalous transport" is used.

6.5   Global Particle Leakage

In addition to the particle-fluid characterization of a plasma, one may identify probabilistic methodologies which, in selected applications are most useful. We discuss here one such approach to the assessment of particle leakage from a fusion reaction volume.

Consider a volume V containing $N_j^*(t)$ particles of the j-type. This population changes with time because of gain and loss reaction rates, R_{+j}^* and R_{-j}^*, respectively, and also because of particle injection and leakage rates, F_{+j}^* and F_{-j}^*, Fig.6.2. The rate equation is evidently

$$\frac{dN_j^*}{dt} = \left(R_{+j}^* - R_{-j}^* \right) + \left(F_{+j}^* - F_{-j}^* \right). \tag{6.43}$$

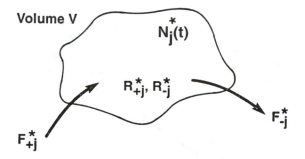

Volume V

$N_j^*(t)$

R_{+j}^*, R_{-j}^*

F_{-j}^*

F_{+j}^*

Fig. 6.2: Volume V containing population $N_j^*(t)$ which is subjected to gain and loss reaction rates, R_{+j}^* and R_{-j}^*, as well as particle injection and leakage rates, F_{+j}^* and F_{-j}^*.

Our interest is specifically in the particle leakage rate F_{-j}^* and we wish to explore an approach which has no need for a detailed space distribution of the particle population; that is, we consider a probabilistic formulation in an integrated global sense for this arbitrary volume.

As our underlying physical basis, we adopt the proposition that the statistically expected fractional change in the population due to particle leakage is $-\Delta N_j^*/N_j^*$ in an interval of time Δt, and is only a function of time. This statement can be expressed in algebraic form as

$$-\left[\frac{\Delta N_j^* / N_j^*}{\Delta t} \right]_{leakage} = \lambda_j(t), \tag{6.44a}$$

where $\lambda_j(t)$ is a positive function of time to be considered subsequently. For now, we take the limit of $\Delta t \to dt$ to write

$$-\left[\frac{1}{N_j^*} \frac{d N_j^*}{dt} \right]_{leakage} = \lambda_j(t). \tag{6.44b}$$

Hence for an initial population $N^*_j(0)$, the fraction of particles remaining up to time t–that is the non-leakage fraction–follows by integration and gives

$$\left[\frac{N^*_j(t)}{N^*_j(0)}\right]_{non\text{-}leakage} = exp\left[-\int_0^t \lambda_j(t')dt'\right]. \tag{6.45}$$

A re-examination of Eq.(6.44b) provides us with a useful expression if we recall that

$$F^*_{-j} = \left|\left[\frac{dN^*_j}{dt}\right]_{leakage}\right| \tag{6.46}$$

and therefore

$$F^*_{-j} = \lambda_j(t) N^*_j(t). \tag{6.47}$$

However, this expression tells us little about the meaning of $\lambda_j(t)$ nor does it provide a suggestion on how to calculate it. This can be resolved if we introduce some probabilistic considerations.

Consider the probability density function $f_j(t)$ which describes the outcome that a particle j remains in the volume V over a time t until it escapes from the ensemble. Hence, this probability is given by the product of the following probabilities which also define $f_j(t)$:

$$f_j(t)dt = \begin{bmatrix} \text{probability of} \\ \text{non - leakage} \\ \text{until time t} \end{bmatrix}\begin{bmatrix} \text{probability of} \\ \text{leakage during} \\ \text{time dt} \end{bmatrix}$$

$$= \left[exp\left(-\int_0^t \lambda_j(t')dt'\right)\right]\left[\frac{|dN^*_j|}{N^*_j}\right]_{dt} \tag{6.48}$$

$$= \left[exp\left(-\int_0^t \lambda_j(t')dt'\right)\right][\lambda_j(t)dt].$$

Note that the substitution here follows directly from Eq.(6.44). Thus, the probability density function of interest is evidently

$$f_j(t) = \lambda_j(t)\, exp\left(-\int_0^t \lambda_j(t')dt'\right). \tag{6.49}$$

with the normalization $\int_0^\infty f_j(t)dt = 1$ as required.

With $f_j(t)$ now specified, we may compute some interesting time dependent quantities. Of particular interest is the particle mean residence time until leakage. Using τ^*_j for this quantity, we may compute it as

$$\tau_j^* = \int_0^\infty t\, f_j(t)\, dt = \int_0^\infty t\left[\lambda_j(t)\, exp\left(-\int_0^\infty \lambda_j(t')dt'\right)\right]dt\,.\qquad(6.50)$$

If the statistical leakage fraction per unit time, Eq.(6.44a), is a constant, then $\lambda_j(t) \to \lambda_j$ and Eq.(6.50) can be integrated to yield

$$\tau_j^* = \int_0^\infty t\, \lambda_j\, exp\left(-\lambda_j t\right)dt = \frac{1}{\lambda_j}\,.\qquad(6.51)$$

That is, λ_j is the inverse mean residence time of the particle in the volume of interest and a more self-explanatory expression for the leakage rate is therefore

$$F_{-j}^* = \frac{N_j^*(t)}{\tau_j^*}\qquad(6.52)$$

where in the context of nuclear fusion, τ_j^* is often called the global particle leakage or confinement time. A good estimate of τ_j^* can frequently be obtained from a knowledge of the mean particle speed and the mean cumulative distance a particle travels in a reaction volume.

    We add that by a similar derivational development, one may show that the energy leakage rate–that is power leakage–from a fusion reaction volume is given by

$$P_-^* = \frac{E^*(t)}{\tau_{E^*}}\qquad(6.53)$$

where $E^*(t)$ is the total energy content of the reaction volume and τ_{E^*} is the global mean energy leakage time or global energy confinement time.

Problems

6.1   Specify the conditions which reduce Eq.(6.21) to its simplest form while still retaining electric and magnetic field effects.

6.2   Consider a particle density $N(x, y, z, E, \theta, \phi)$ and formulate a reduced transport equation, Eq.(6.42), for each of the following cases:
    (a) neutral particles,
    (b) monoenergetic particles,
    (c) only the z-direction is relevant, and
    (d) both the source and collision terms are given.

6.3   Consider a plasma contained between two walls x_o units apart and effectively infinite in the y and z-directions. For the case that the plasma decays by diffusion to the walls and for separability of the form $N(\mathbf{r},t) = R(\mathbf{r})\, T(t)$, determine $N(\mathbf{r},t)$ throughout this plasma slab region for all time. Sketch this density function.

6.4   Confirm that for a cylindrically-shaped, magnetically confined plasma wherein radial diffusion is the dominant mode of transport, assuming a constant volumetric ion source, the steady state ion density will have the following relationship:

$$N(r) = N(0)\sqrt{1 - \frac{r^2}{a^2}} \, , \qquad\qquad (6.54)$$

where $N(0)$ is the ion density at $r = 0$, and a is the plasma minor radius. (Note that for these conditions, Eq.(6.10b) becomes $\nabla \times \mathbf{J}(\mathbf{r}) = S = \text{constant}$.)

6.5   Given Eq.(6.54) for the ion density radial distribution, derive the average ion density.

7.   Fusion Burn

The purpose of a fusion energy device is to initiate and sustain nuclear fusion reactions under acceptable operational conditions. An understanding of the reaction rates and the time dependence of burn processes is therefore of importance and will now be considered.

7.1   Elementary D-T Burn

Consider a deuterium-tritium ion population at thermonuclear temperatures in a suitable unit volume of interest. Regardless of what approach is used to heat and confine the ions, a quantity of energy Q_{dt} is released whenever the event

$$d + t \rightarrow n + \alpha \qquad (7.1)$$

occurs. As discussed in Sec. 2.5, these reactions proceed at the rate

$$R_{dt}(\mathbf{r},t) = \iint N_d(\mathbf{r},\mathbf{v}_d,t) N_t(\mathbf{r},\mathbf{v}_t,t) \sigma_{dt}\left(\left|\mathbf{v}_d - \mathbf{v}_t\right|\right)\left|\mathbf{v}_d - \mathbf{v}_t\right| d^3\mathbf{v}_d\, d^3\mathbf{v}_t$$

$$= N_d^* N_t^* \iint f_d(\mathbf{r},\mathbf{v}_d,t) f_t(\mathbf{r},\mathbf{v}_t,t) \sigma_{dt}\left(\left|\mathbf{v}_d - \mathbf{v}_t\right|\right)\left|\mathbf{v}_d - \mathbf{v}_t\right| d^3\mathbf{v}_d\, d^3\mathbf{v}_t$$

$$= N_d(\mathbf{r},t) N_t(\mathbf{r},t)\frac{\iint f_d(\mathbf{r},\mathbf{v}_d,t) f_t(\mathbf{r},\mathbf{v}_t,t) \sigma_{dt}\left(\left|\mathbf{v}_d - \mathbf{v}_t\right|\right)\left|\mathbf{v}_d - \mathbf{v}_t\right| d^3\mathbf{v}_d\, d^3\mathbf{v}_t}{\iint f_d(\mathbf{r},\mathbf{v}_d,t) f_t(\mathbf{r},\mathbf{v}_t,t) d^3\mathbf{v}_d\, d^3\mathbf{v}_t}$$

$$= N_d(\mathbf{r},t) N_t(\mathbf{r},t) <\sigma v>_{dt}(\mathbf{r},t),$$

$$(7.2a)$$

in a unit volume where use has been made of Eqs.(6.40) and (6.41). Note that $<\sigma v>_{dt}$ appears correctly now as a function of position and time, arising from the averaging process over the particle distribution functions which explicitly depend upon these variables. Accordingly, the rate of energy-density release–that is the fusion power density–is given by

$$P_{dt}(\mathbf{r},t) = R_{dt}(\mathbf{r},t) Q_{dt} = N_d(\mathbf{r},t) N_t(\mathbf{r},t) <\sigma v>_{dt}(\mathbf{r},t) Q_{dt} . \qquad (7.2b)$$

Recall that Q_{dt} = 17.6 MeV and $<\sigma v>_{dt}$ is, of course, primarily a function of the ion temperature and is graphically displayed in Fig.2.5 with a tabulation listed in Appendix C.

Evidently then, the fusion power in a unit volume will change if the ion temperature changes with time or if the fusile ion densities vary with time. These variations are governed by the occurring reactions, by the injection rates, and by the leakage rates, all of which add to or subtract from the particle and energy densities. Adopting our previous rate equation analysis of the form of Eq.(6.4) to that of particle density and energy density variations at a fixed point in space, i.e., $r \neq r(t)$–or for the case of uniform ion distributions–so that total derivatives can be used, we choose as an appropriate description for a particular ion density $N_j(r,t)$ the general rate equation

$$\frac{dN_j}{dt} = \left(F_{+j} - F_{-j}\right) + \left(R_{+j} - R_{-j}\right). \tag{7.3}$$

Here, as well as in the following, we omit the space-dependent notation in view of the fact that we are only considering the time dependence of the particle density. Further, F_{+j} and F_{-j} are the ion injection rate and leakage rate densities, with F_{-j} equivalent to the term $\nabla \times J_j(r,t)$ in the Continuity Equation, Eq.(6.10b), and R_{+j} is the reaction rate density which adds to $N_j(t)$ while R_{-j} is the reaction rate density which removes particles from the population $N_j(t)$ in the fixed unit volume.

For the special case of zero ion leakage and no ion injection, or if these two processes are exactly equal and hence cancel each other in Eq.(7.3), and for $R_{+j}=0$, the instantaneous time dependence of the deuterium and tritium ions' densities is therefore specified by the rate equations

$$\frac{dN_d}{dt} = - R_{dt}(t) = - N_d(t) N_t(t) < \sigma v >_{dt} \tag{7.4a}$$

and similarly

$$\frac{dN_t}{dt} = - N_d(t) N_t(t) < \sigma v >_{dt} . \tag{7.4b}$$

These equations can be combined in the form

$$\frac{d}{dt}\left(N_d + N_t\right) = -2 N_d(t) N_t(t) < \sigma v >_{dt} . \tag{7.5}$$

Suppose that a 50:50% ratio between the deuterium and tritium ions exists so that

$$N_i(t) = N_d(t) + N_t(t) = \tfrac{1}{2} N_i(t) + \tfrac{1}{2} N_i(t) \tag{7.6}$$

is the total fuel ion density at an arbitrary time during the burn; this gives for Eq.(7.5) therefore

$$\frac{dN_i}{dt} = - \frac{< \sigma v >_{dt}}{2} N_i^2(t) . \tag{7.7}$$

This differential equation is valid in the unit volume of interest and for a burn time under conditions of $F_{+i} - F_{-i} = 0$. Burn times can be very short, of the order of 10^{-8} s for inertially confined fusion, or of the order of seconds–and envisaged to be much longer–for magnetically confined systems.

The associated instantaneous fusion power density generated in this burn is obtained by multiplying the d-t reaction rate density with the energy release Q_{dt} per fusion event and thus

$$P_{dt}(t) = N_d(t) N_t(t) <\sigma v>_{dt} Q_{dt} = \frac{<\sigma v>_{dt} Q_{dt}}{4} N_i^2(t). \tag{7.8}$$

A determination of the time dependence of the burn requires a knowledge of how the fuel density $N_i(t)$ at a fixed position in space varies during a burn time τ_b. This variation is described by Eq.(7.7) for the case $F_{+i} - F_{-i} = 0$ and $R_{+i} = 0$, and represents a differential equation which is separable to give

$$\int_{N_{i,o}}^{N_i(t)} \frac{dN_i}{N_i^2} = -\frac{1}{2} \int_0^t <\sigma v>_{dt} dt \tag{7.9}$$

for $0 \le t \le \tau_b$, with $N_i(0) = N_{i,o}$ as an initial condition. The left part can be integrated so that

$$-\frac{1}{N_i(t)} + \frac{1}{N_{i,o}} = -\frac{1}{2} \int_0^t <\sigma v>_{dt} dt \tag{7.10a}$$

or

$$N_i(t) = \frac{1}{\dfrac{1}{N_{i,o}} + \dfrac{1}{2}\int_0^t <\sigma v>_{dt} dt}. \tag{7.10b}$$

If the ion temperature remains sufficiently constant during the burn–implying that, for example, the alpha particle reaction products transfer some of their kinetic energy to the plasma and so compensate for energy loss by bremsstrahlung radiation, or that external heating is supplied or that excess energy is removed–then $<\sigma v>_{dt}$ is also a constant and the resultant isothermal fuel ion density variation with time is simply

$$N_i(t) = \frac{1}{\dfrac{1}{N_{i,o}} + \dfrac{1}{2} <\sigma v>_{dt} t}. \tag{7.11}$$

The fuel burnup fraction is then found to be given by the compact expression

$$f_b = \frac{N_{i,o} - N_i(\tau_b)}{N_{i,o}} = \frac{1}{1 + \left[\dfrac{N_{i,o}}{2} <\sigma v>_{dt} \tau_b\right]^{-1}}. \tag{7.12}$$

Thus, the important consequence is that the burnup fraction is determined by the product of three factors: $N_{i,o}$, $<\sigma v>_{dt}$, and τ_b. Figure 7.1 illustrates this variation of the burn fraction with the ion temperature T for several burn times.

The fusion power released in a fixed unit volume, Eq.(7.8), is now given by

$$P_{dt}(t) = \frac{N_i^2(t)}{4} <\sigma v>_{dt} Q_{dt} = \frac{1}{4}\left[\frac{1}{\dfrac{1}{N_{i,o}} + \dfrac{1}{2} <\sigma v>_{dt} t}\right]^2 <\sigma v>_{dt} Q_{dt} \tag{7.13}$$

and evidently tends to decrease more rapidly with time than the fuel ion density, Fig.7.2.

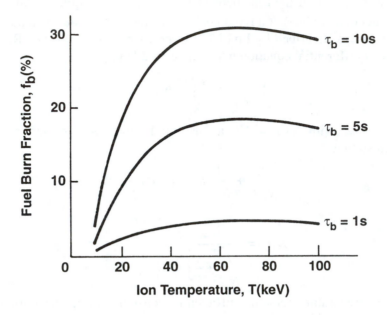

Fig. 7.1: Fusion fuel burn fraction as a function of ion temperature and burn time for d-t fusion. The initial ion density inventory has been taken as $N_{i,o} = 10^{20}$ m^{-3}.

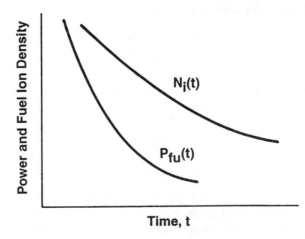

Fig. 7.2: Dominant variation of the power and fuel ion density with time.

Finally, the total energy released in a unit volume at a fixed **r** during the isothermal fusion burn is given by

$$E_{dt} = \int_0^{\tau_b} P_{dt}(t)dt = \frac{Q_{dt}}{4} <\sigma v>_{dt} \int_0^{\tau_b} \left(\frac{1}{\frac{1}{N_{i,o}} + \frac{1}{2} <\sigma v>_{dt} t} \right)^2 dt \ . \tag{7.14}$$

In general, with the ion temperature varying during the burn time, the more exact non-isothermal expression can be shown to be

$$E_{dt} = \frac{Q_{dt}}{4} \int_0^{\tau_b} \frac{<\sigma v>_{dt}(t)}{\left(\frac{1}{N_{i,o}} + \frac{1}{2} \int_0^t <\sigma v>_{dt} dt' \right)^2} dt \ . \tag{7.15}$$

7.2 Comprehensive D-T Burn

A more comprehensive d-t isothermal burn analysis needs to include not only the destruction of deuterium and tritium by fusion reactions but also the injection and leakage rates of ions into and from the unit volume of interest. That is, the rate density equation, Eq.(7.3), referring to a fixed point in space is now given by the following:

$$\frac{dN_j}{dt} = \left(\frac{\partial N_j}{\partial t} \right)_{in} - \left(\frac{\partial N_j}{\partial t} \right)_{leak} - \left(\frac{\partial N_j}{\partial t} \right)_{burn} \ . \tag{7.16}$$

The ion injection rate density, which is a reactor operations function, will be represented by the fueling rate density

$$F_j = \left(\frac{\partial N_j}{\partial t} \right)_{in} \tag{7.17}$$

where here, and in the following, we omit the + in front of the j. Ion leakage from a unit volume may be taken to be proportional to the ion density so that

$$\left(\frac{\partial N_j}{\partial t} \right)_{leak} = \frac{N_j(t)}{\tau_j} \tag{7.18}$$

where τ_j is a characteristic time parameter representing the local mean residence time of the ions. The determination of this time quantity in terms of various reactor parameters is a continuing task of fusion research but for present purposes here, τ_j is taken as a known parameter. Recalling our previous analysis leading to a global particle confinement time and noting that

$$\frac{1}{\tau_j^*} = \frac{F_j^*}{N_j^*} = \frac{1}{N_j^*} \int_V \frac{N_j(\mathbf{r},t)}{\tau_j(\mathbf{r},t)} d^3r = \frac{1}{\tau_j} \tag{7.19}$$

suggests thereby that the global particle confinement time is the density-weighted volume average of the local mean residence time. Both these residence times generally vary with time.

The more complete fuel balance equations for d-t burn are therefore given by the coupled set of equations

$$\frac{dN_d}{dt} = F_d - N_d(t)\, N_t(t) <\sigma v>_{dt} - \frac{N_d(t)}{\tau_d}\,, \tag{7.20a}$$

$$\frac{dN_t}{dt} = F_t - N_d(t)\, N_t(t) <\sigma v>_{dt} - \frac{N_t(t)}{\tau_t} \tag{7.20b}$$

or combining, as was done for Eq.(7.5),

$$\frac{d}{dt}(N_d + N_t) = (F_d + F_t) - 2\, N_d(t)\, N_t(t) <\sigma v>_{dt} - \left(\frac{N_d(t)}{\tau_d} + \frac{N_t(t)}{\tau_t}\right). \tag{7.20c}$$

An equal mixture of deuterium and tritium with identical injection and leakage rates is specified by the conditions

$$N_i = N_d(t) + N_t(t) = \frac{N_i(t)}{2} + \frac{N_i(t)}{2}\,, \tag{7.21}$$

$$F_i = F_d + F_t = \frac{F_i}{2} + \frac{F_i}{2}\,, \tag{7.22}$$

$$\tau_i = \tau_d + \tau_t \tag{7.23}$$

where the subscript i refers to all fuel ions. This then leads to the dynamical equation for the fuel ion density at a fixed position in space and at some time t, for the three processes here identified, as

$$\frac{dN_i}{dt} = F_i - \frac{N_i(t)}{\tau_i} - \frac{<\sigma v>_{dt}}{2}\, N_i^2(t), \qquad 0 \le t \le \tau_b\,. \tag{7.24}$$

This is a nonlinear first order differential equation, frequently called Ricatti's equation, and like many nonlinear equations in general, possesses some subtle features. However, by combining several analytical, geometrical and physical notions, we can extract some useful qualitative information about the time variation of the fuel ion density $N_i(t)$ at a fixed position in space and hence about the power density $P_{dt}(t)$, Eq.(7.2b).

For algebraic simplicity, we write Eq.(7.24) as

$$\frac{dN}{dt} = a_0 + a_1 N + a_2 N^2 \tag{7.25}$$

with $a_0 = F_i$, $a_1 = -1/\tau_i$ and $a_2 = -<\sigma v>_{dt}/2$. For the special case when these parameters are constant, then the following information about the slope of an N vs. t plot at an arbitrary time can be specified:

$$\text{slope } (= a_0 + a_1 N + a_2 N^2) = \begin{cases} \approx a_0 > 0 & \text{for sufficiently small N} \\ \approx a_2 N^2 < 0 & \text{for sufficiently large N}\,. \end{cases} \tag{7.26}$$

Further, for the special case of a_i constant, any fixed points of Eq.(7.25)–to be represented by N^o–are the solutions of $dN/dt = 0$; that is

$$a_0 + a_1 N^o + a_2 (N^o)^2 = 0 \tag{7.27}$$

yielding two distinct roots

$$N_{\pm}^o = \frac{-a_1 \pm \sqrt{a_1^2 - 4 a_0 a_2}}{2 a_2} \,.$$ (7.28a)

We take the negative root–which yields a positive density–and substitute the original physical parameters to obtain

$$N_{i-}^o = \frac{-1/\tau_i + \sqrt{1/\tau_i^2 + 2 F_i <\sigma v>_{dt}}}{<\sigma v>_{dt}} \,.$$ (7.28b)

The above information about the slope of $N(t)$ for small and large N's, and the recognition that $N(t) \rightarrow N^o$ for $t \rightarrow \infty$–and the assumption that a_0, a_1 and a_2 can be adequately represented as suitable constants in time–allows us to sketch the solution of $N_i(t)$ of Eq.(7.24) as depicted in Fig.7.3.

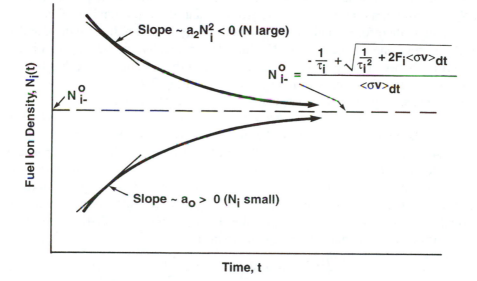

Fig. 7.3: Representation of the fuel ion density as a function of time for the case of a quadratic autonomous dynamic equation $dN/dt = a_0 + a_1 N + a_2 N^2$.

7.3   Identical Particle Burn

While the first generation of fusion reactors will evidently be d-t fueled, it is expected that subsequent generations of fusion reactors may use pure deuterium and possibly other "advanced" fuels as discussed in Sec.1.4; one advantage thus gained is the elimination of the need to breed radioactive tritium. However, even though tritium breeding is not necessary, tritium handling will nevertheless still be required because it is one of the reaction products of d-d fusion, Eq.(1.21).

In comparison to d-t fusion, d-d fusion represents reactions among identical particles. The question of interest is therefore the following: given that for distinguishable ions, we have

$$a + b \rightarrow (\ ) + (\ ) \tag{7.29}$$

with

$$R_{ab} = N_a N_b < \sigma v >_{ab}, \tag{7.30}$$

what is the corresponding reaction rate expression for

$$a + a \rightarrow (\ ) + (\ ) \tag{7.31}$$

for the case of arbitrary reaction products? That is, what is R_{aa} for this latter case? These considerations are important not only for d-d but also for t-t and h-h fusion.

Some thought suggests that the product $N_a N_b$ in Eq.(7.29) represents the number of ways that members of the a-type set of particles can combine with members of the b-type set. As suggested in Fig.7.4, any one of the particles from the N_a set could combine with any one from N_b and the total number of possible interactions is therefore equal to the number of (x,y) combinations with x and y identifying any one of the N_a and N_b particles, respectively. Hence, all binary possibilities between these two sets are included in the product $N_a N_b$; that is, this product represents the totality of matrix elements in Fig.7.4.

For the case of particle indistinguishability, however, we have

$$a = b. \tag{7.32}$$

Thus, with the total reactant density being N_i and as suggested by the extension of Fig.7.4, it is necessary to exclude the diagonal elements because a particle cannot interact with itself. However, the remaining $N_i^2 - N_i$ combinations–that is the total number of matrix elements minus the number of diagonal terms– includes also the transposed pairs (x,y) and (y,x) which clearly represent the identical process and event. Hence, the total number of distinct reaction possibilities is only one half of this, giving therefore

$$\frac{N_i^2 - N_i}{2} = \frac{1}{2} N_i (N_i - 1) \approx \frac{N_i^2}{2}. \tag{7.33}$$

The last approximation is, of course, essentially exact because $N_i \gg 1$ for all cases of general interest.

We write therefore for the reaction rate density among identical particles

$$R_{aa} = < \sigma v >_{aa} \frac{N_a^2}{2} \tag{7.34}$$

where N_a is the total number of indistinguishable ions per unit volume; this number differs by a factor of 2 from the case of distinguishable fusing species, Eq.(7.30).

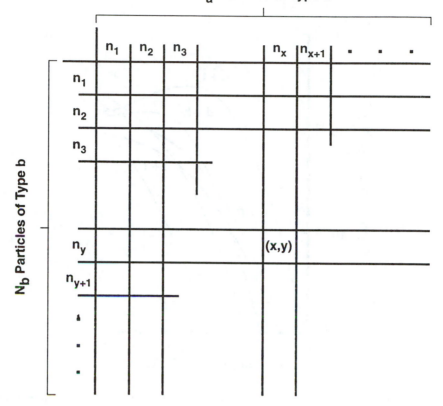

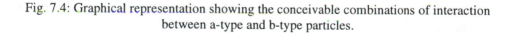

Fig. 7.4: Graphical representation showing the conceivable combinations of interaction between a-type and b-type particles.

7.4 D-D Burn Modes

The use of only deuterium as a fusion fuel introduces several important considerations. For temperatures up to about 200 keV, the d-d reaction parameter sigma-v is substantially smaller (roughly two orders of magnitude) than it is for d-t fusion, Fig.7.5; hence, for equal particle densities and ion temperatures below about 200 keV, a d-d fusion reactor will possess a much smaller power density and hence will require a larger size for a specified total fusion power production. Note additionally that radiation losses tend to increase with higher temperature, thereby further contributing to problems of a viable power balance. Nevertheless, deuterium based fusion possesses some very appealing properties. For example, an important feature of the deuterium fueled reactor is that the products from d-d fusion may fuse with the deuterium fuel and possibly even among themselves.

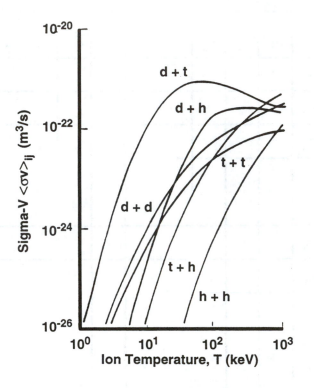

Fig. 7.5: Sigma-v for various fusing fuels characterized by Maxwellian ion distributions.

The primary d-d fusion reaction proceeds via two (almost) equally likely reaction channels

$$d + d \rightarrow \begin{cases} t + p \\ h + n \end{cases}$$ (7.35)

where for notational simplicity, we use $h = {}^3He$. The bred tritium and helium-3 nuclei possess significant sigma-v parameters to fuse with the deuterium fuel, Fig.7.5. The reaction

$$d + t \rightarrow \alpha + n$$ (7.36)

will be dominant with

$$d + h \rightarrow \alpha + p$$ (7.37)

also taking place. In addition, there also exists the possibility of the reaction products to fuse among themselves to yield

$$t + t \rightarrow 2n + \alpha$$ (7.38a)

$$h + h \rightarrow 2p + \alpha$$ (7.38b)

$$t + h \rightarrow \begin{cases} p + n + \alpha \\ d + \alpha . \end{cases} \qquad (7.38c)$$

Reactions involving reaction products are often called side reactions.

A complex system of linked reactions may possibly emerge. The following suggests a classification of d-d sustained reaction systems.

(a) PURE-D Mode

The idealized case of d-d fusion only is given by

$$d + d \rightarrow t + p \quad \text{(channel - t)} \qquad (7.39a)$$
$$d + d \rightarrow h + n \quad \text{(channel - h)} \qquad (7.39b)$$

which proceed at the rates

$$R_{dd,t} = \frac{N_d^2}{2} <\sigma v>_{dd,t} , \quad Q_{dd,t} = 4.1 \text{ MeV} \qquad (7.40a)$$

$$R_{dd,h} = \frac{N_d^2}{2} <\sigma v>_{dd,h} , \quad Q_{dd,h} = 3.2 \text{ MeV} \qquad (7.40b)$$

where the reaction Q-values are extracted from Table 7.1. Here, it is also essential to introduce the additional t and h subscript notation according to the channel designations of Eq.(7.39); values for $<\sigma v>_{dd,t}$ and $<\sigma v>_{dd,h}$ are listed in Table C.1 of Appendix C.

Further,

$$<\sigma v>_{dd} = <\sigma v>_{dd,t} + <\sigma v>_{dd,h} \qquad (7.41a)$$

for which, to a very good approximation

$$<\sigma v>_{dd,t} \approx <\sigma v>_{dd,h} \approx \frac{1}{2} <\sigma v>_{dd} \qquad (7.41b)$$

at temperatures of common interest.

(b) SCAT-D Mode

The very large $<\sigma v>_{dt}$ parameter, Fig.7.5, suggests that for most of the temperature range shown, the bred tritium will be consumed almost immediately upon production while the bred helium-3 will not be burned so rapidly due to the smaller $<\sigma v>_{dh}$ at these temperatures. This fusion reaction mode may therefore be represented as

$$d + d \rightarrow t + p \quad \text{(channel - t)} \qquad (7.42a)$$
$$\searrow$$
$$d + t \rightarrow n + \alpha \qquad (7.42b)$$
$$d + d \rightarrow h + n \quad \text{(channel - h)} . \qquad (7.42c)$$

Here the arrow suggests a reaction link. A summary reaction representation for this cycle is

$$5d \rightarrow 2n + h + \alpha + p , \quad Q_{SCAT-D} = 24.9 \text{ MeV} \qquad (7.43)$$

providing Eqs.(7.42a) and (7.42b) occur at equal rates, i.e. $R_{dd,t} = R_{dt}$ so that

Fusion Reactions (Particle energies in brackets, in MeV)	Reaction Q-Value (MeV)	Charged Particle Energy Fraction	Neutron Energy Fraction
D + T Burn			
d + t → n (14.1) + α (3.5)	17.6	0.20	0.80
D + D Burn			
PURE-D			
d + d → t (1.0) + p (3.1)	4.1		
d + d → h (0.8) + n (2.4)	3.2		
4d　→ h + n + t + p	7.3	0.67	0.33
SCAT-D Burn			
d + d → t (1.0) + p (3.1)	4.1		
d + t → n (14.1) + α (3.5)	17.6		
d + d → h (0.8) + n (2.4)	3.2		
5d　→ h + 2n + α + p	24.9	0.34	0.66
CAT-D Burn			
d + d → t (1.0) + p (3.1)	4.1		
d + t → n (14.1) + α (3.5)	17.6		
d + d → h (0.8) + n (2.4)	3.2		
d + h → α (3.7) + p (14.6)	18.3		
6d　→ 2n + 2α + 2p	43.2	0.62	0.38

Table 7.1: Tabulation of various hydrogen-based fusion burn modes and associated energy data.

$$\frac{N_d^2}{2} <\sigma v>_{dd,t} = N_d N_t <\sigma v>_{dt} \ .$$

$$\tag{7.44}$$

This relationship implies

$$\frac{N_t}{N_d} = \frac{1}{2} \frac{<\sigma v>_{dd,t}}{<\sigma v>_{dt}} \approx \frac{1}{4} \frac{<\sigma v>_{dd}}{<\sigma v>_{dt}}$$

$$\tag{7.45}$$

and thus provides for triton fusion burn at a rate equal to its production rate. This fusion operation mode is often called the "semi-catalyzed-D cycle" (SCAT-D). By reference to Fig.7.5, the relative tritium concentration in the fusing plasma may therefore be small at low-to-medium temperatures but will increase for higher temperatures.

(c) CAT-D Mode

An examination of Fig.7.5 suggests that the fusing of helium-3 nuclei with deuterons is the next most likely nuclear reaction. This completes the catalysing process, hence the name CAT (catalyzed) cycle:

$$d + d \rightarrow t + p \quad \text{(channel - t)} \tag{7.46a}$$

$$d + t \rightarrow n + \alpha \tag{7.46b}$$

$$d + d \rightarrow h + n \quad \text{(channel - h)} \tag{7.46c}$$

$$d + h \rightarrow \alpha + p \tag{7.46d}$$

which is equivalent to

$$6d \rightarrow 2n + 2\alpha + 2p, \quad Q_{CAT-D} = 43.2 \text{ MeV} \tag{7.47}$$

provided that the four reactions proceed at equal rates.

The exact sustainment of only one of the above specific d-d burn modes may in general be very difficult. The more general case is suggested in Table 7.2 where the connection reaction linkages will evidently vary with temperature and density.

7.5   D-^{3}He Fusion

We had previously noted that d-t fusion involves radioactive tritium as a fuel and high energy neutrons as reaction products; the former poses problems of radiological safety because tritium diffuses readily, while the latter leads to difficulties of first-wall endurance, shielding and induced radioactivity. Similarly d-d fusion produces both tritons and neutrons, e.g. Table 7.1.

From an examination of the temperature dependence of sigma-v, Fig.7.5, we note that of the five most readily attainable fusion reactions, only

$$d + h \rightarrow p + \alpha, \quad Q_{dh} = 18.3 \text{ MeV} \tag{7.48}$$

does not involve neutrons or tritons. For this reason, d-h is often called an attainable "clean" fusion reaction.

As suggested in Fig.7.5, the maximum $<\sigma v>_{dh}$ occurs at temperatures higher than the maximum for d-t fusion, requiring therefore higher reaction temperatures. In addition, at higher temperatures–as well as for reasons of a higher proton number in helium than in hydrogen–bremsstrahlung radiation is

$d + d \rightarrow t\ (1.0) + p\ (3.1)$

$d + t \rightarrow n\ (14.1) + \alpha\ (3.5)$

$t + t \rightarrow n\ (5.0) + n\ (5.0) + \alpha\ (1.3)$

$t + h \rightarrow p\ (5.7) + \alpha\ (1.3) + n\ (5.1)$

$d + d \rightarrow h\ (0.8) + n\ (2.4)$

$d + h \rightarrow \alpha\ (3.7) + p\ (14.6)$

$h + h \rightarrow p\ (5.7) + p\ (5.7) + \alpha\ (1.4)$

Table 7.2: Characteristics of the general d-d initiated fusion linkage processes. (Particle Energies in brackets, in MeV.)

more severe.

Further, there exists the question of an adequate helium-3 supply. It is known that ^{3}He is scarce, occurring with a natural abundance of ^{3}He/(^{3}He + ^{4}He) $\approx 10^{-6}$. An existing supply of tritium, however, eventually produces helium-3 by nuclear decay

$$t \rightarrow h + \beta^- , \quad \tau_{1/2} = 12.3 \text{ years} . \tag{7.49}$$

with $\tau_{1/2}$ denoting the half-life of tritium. In addition, one may conceive of a d-d fusion reaction so that the helium-3 produced via

$$d + d \rightarrow n + h \tag{7.50}$$

could serve as a helium-3 fuel source. However, a most significant source of this terrestrially scarce fusion fuel has recently been identified in lunar rock samples. It appears therefore that ^{3}He could be mined in sufficient quantity on the moon's surface and transported to the earth under energetically favourable conditions of reaction (7.48).

The designation of d-h as a clean reaction needs to be somewhat tempered because of the potential for unclean side reactions. That is, while the principal reaction proceeds according to

$$d + h \rightarrow p + \alpha, \quad R_{dh} = <\sigma v>_{dh} N_d N_h , \tag{7.51}$$

the existence of a deuteron population N_d will evidently enable the following triton and neutron producing reactions to occur simultaneously:

$$d + d \rightarrow t + p, \quad R_{dd,t} = <\sigma v>_{dd,t} \frac{N_d^2}{2}$$

(7.52a)

$$d + d \rightarrow h + n, \quad R_{dd,h} = <\sigma v>_{dd,h} \frac{N_d^2}{2}.$$

(7.52b)

Increasing the number of desirable d-h reactions relative to the undesirable triton and neutron production rates is thus an important objective for "cleanliness". These ratios are evidently

$$\frac{R_{dh}}{R_{dd,t}} = 2 \frac{<\sigma v>_{dh}}{<\sigma v>_{dd,t}} \frac{N_h}{N_d}$$

(7.53a)

and

$$\frac{R_{dh}}{R_{dd,h}} = 2 \frac{<\sigma v>_{dh}}{<\sigma v>_{dd,h}} \frac{N_h}{N_d}$$

(7.53b)

and point to the importance of a careful specification of the temperature and the role of the helium-to-deuterium ion population.

These two conclusions provide some hints for the type of fusion cycle one should seek if a high degree of "cleanliness" is to be attained: good control on high temperature and good control of the helium-3 and deuterium fuel ions.

7.6   Spin Polarized Fusion

Our discussion of fusion burn thus far would seem to suggest that only simple collision considerations are relevant. However, fusion phenomena may include other intrinsic processes thereby increasing the fusion reaction rate, and a possible suppression of less-desirable side reactions. We consider one such process next.

Neutrons and protons are known to possess an intrinsic spin. Quantum mechanical considerations characterize this spin as $\frac{1}{2}\hbar$ with $\hbar$ representing Planck's constant. Then, when nucleons combine to form nuclei, the constituent spins add vectorially assigning a spin to the entire nucleus. The nuclides involved in d-t fusion possess spins as follows in units of the Planck constant $\hbar$:

deuteron : 1
triton   : ½
neutron : ½
alpha   : 0 .

The direction of a nuclear spin is determined by reference to an external magnetic field with the convention that the (+) sign represents a direction parallel to the magnetic field, (-) refers to the anti-parallel direction and (0) is for the transverse direction. We suggest these spins for a deuteron and a triton in Fig.7.6 where we also introduce the notation $f^0{}_0$ for the fraction of the respective nuclei oriented in the allowed directions; quantum mechanical considerations permit no other directions.

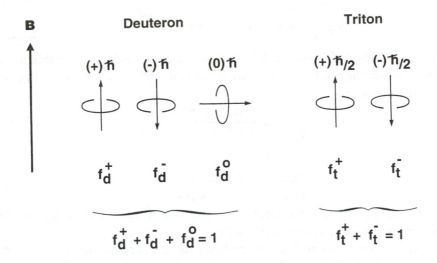

Fig. 7.6: Spins for deuterons and tritons.

It is known that spin conservation characterizes nuclear transformations. The maximum spin for ^{5}He, produced momentarily as a compound nucleus when the reaction

$$d + t \rightarrow \left(^5 He\right)^* \rightarrow n + \alpha \tag{7.54}$$

occurs, is 3/2 in units of $\hbar$—occurring when the deuteron spin and triton spin are aligned. The microscopic fusion cross section for this process will be represented by σ_o and constitutes a maximum.

It is possible to write an expression for the d-t fusion cross section σ_{dt} in terms of σ_o and the various fractional concentrations of deuterons and tritons in their allowed spin states, i.e., in terms of f_d^+, f_d^-, f_d^o, f_t^+, and f_t^- as depicted in Fig.7.6. Referring to some specialized aspects of nuclear reaction theory, this cross section can be well represented by

$$\sigma_{dt} = \left[\left(f_d^+ f_t^+ + f_d^- f_t^- \right) + \tfrac{2}{3} \left(f_d^o \right) + \tfrac{1}{3} \left(f_d^+ f_t^- + f_d^- f_t^+ \right) \right] \sigma_o \tag{7.55}$$

and can be used to evaluate the extent to which σ_{dt} approaches the maximum possible fusion cross-section, σ_o, for various mixtures of polarization of deuterons and tritons in a magnetic field.

Evidently, an equal fraction of fuel nuclei in each of their allowed spin states constitutes an overall completely depolarized state; such random polarization is defined by the ratios

$$f_d^+ = \tfrac{1}{3}, \quad f_d^- = \tfrac{1}{3}, \quad f_d^o = \tfrac{1}{3}, \quad f_t^+ = \tfrac{1}{2}, \quad f_t^- = \tfrac{1}{2} \tag{7.56}$$

and the corresponding d-t fusion cross section $(\sigma_{dt})_{r,r}$ is therefore given by

$$(\sigma_{dt})_{r,r} = \left\{ \left[\left(\tfrac{1}{3}\right)\left(\tfrac{1}{4}\right) + \left(\tfrac{1}{3}\right)\left(\tfrac{1}{2}\right) \right] + \tfrac{2}{3}\left(\tfrac{1}{3}\right) + \tfrac{1}{3}\left[\left(\tfrac{1}{3}\right)\left(\tfrac{1}{4}\right) + \left(\tfrac{1}{3}\right)\left(\tfrac{1}{2}\right) \right] \right\} \sigma_o = \tfrac{2}{3}\sigma_o \ . \tag{7.57}$$

Thus, for this case of random spin polarization, the σ_{dt} cross section equals 2/3 of the maximum possible value, σ_o.

Consider now a parallel alignment of the deuterium and tritium nuclei. For this case we have

$$f_d^+ = 1, \quad f_d^- = 0, \quad f_d^o = 0, \quad f_t^+ = 1, \quad f_t^- = 0 \tag{7.58}$$

and hence

$$(\sigma_{dt})_{+,+} = (1 + 0 + 0 + 0 + 0)\sigma_o = \sigma_o \ . \tag{7.59}$$

Comparison of this result with Eq.(7.57) indicates that a 50% increase in the d-t fusion reaction rate density is achieved if the deuterons and tritons all possess a spin alignment in the direction of the magnetic field. By inspection, the same result occurs if the spins all point in the opposite direction, i.e.,

$$(\sigma_{dt})_{+,+} = (\sigma_{dt})_{-,-} \ . \tag{7.60}$$

A further question of interest then is to consider a plasma for which the tritons spins are random and the deuterons are injected with a specified spin polarization. Setting therefore

$$f_d^+ = 1, \quad f_d^- = 0, \quad f_d^o = 0, \quad f_t^+ = \tfrac{1}{2}, \quad f_t^- = \tfrac{1}{2} \tag{7.61}$$

gives

$$(\sigma_{dt})_{+,r} = \left[\tfrac{1}{2} + 0 + 0 + \tfrac{1}{3}\left(\tfrac{1}{2}\right) + 0 \right]\sigma_o = \tfrac{2}{3}\sigma_o \tag{7.62}$$

as for Eq.(7.57).

Accelerators exist which could supply spin polarized deuterium or tritium ions into a plasma. Then, since the spin polarization can be sustained for time periods up to 10 s, a 50% increased fusion power density could be sustained. Additionally, it has been found that under conditions of polarized fusion, the neutron and alpha reaction products emerge with a preferred directional distribution requiring therefore special containment-wall considerations.

7.7   Catalyzed Fusion

A reaction statement for d-t fusion which incorporates both mass flow and energy flow may be written as

$$E_{dt} + d + t \rightarrow (dt)^* \rightarrow n + \alpha + Q_{dt} \tag{7.63}$$

where E_{dt} is the energy supplied to heat the deuterium and tritium to thermonuclear temperatures and to sustain their high kinetic energies against all energy loss mechanisms; $(dt)^*$ is the short-lived intermediate nuclear state previously denoted by $(^5He)^*$. Energy viability evidently demands $E_{dt} < Q_{dt}$.

In establishing fusion reactor conditions it is the electrostatic repulsion between the fuel ions which demands intensive heating of the plasma to high temperatures in order that the Coulomb barrier between two fusing nuclei can be more easily penetrated or overcome providing for a substantial fusion reaction rate. One might therefore speculate about the appealing prospects of using some "agent x" which could neutralize the Coulomb repulsion and induce a fusion event by making it easier for the ions to enter the range of their strong nuclear forces of attraction. That is, this agent might serve the function of a "catalyst" similar to the common practice of using catalyzing compounds to enhance the rate of chemical reactions. In particular, this catalytic agent x must be released after the d-t fusion event and subsequently induce another fusion reaction. Taking the average energy cost of producing one agent x to be E_x and if each x catalyzes χ fusion events, then we write for the entire sequence

$$E_x + d + t + x \rightarrow (xdt)^* \rightarrow x + n + \alpha + Q_{dt}$$

$$d + t + x \rightarrow (xdt)^* \rightarrow x + n + \alpha + Q_{dt}$$

$$d + t + x \rightarrow (xdt)^* \rightarrow x + n + \alpha + Q_{dt} \qquad (7.64)$$

$$\cdot \qquad \cdot \qquad \cdot \qquad \cdot$$
$$\cdot \qquad \cdot \qquad \cdot \qquad \cdot$$
$$\cdot \qquad \cdot \qquad \cdot \qquad \cdot$$

$$\chi \, Q_{dt}$$

Energy viability now requires

$$E_x < \chi Q_{dt} . \qquad (7.65)$$

With Q_{dt} a known constant, the fusion catalyst should possess an average energy cost of production E_x and catalyze χ d-t events such that now Eq.(7.65) is satisfied.

Interestingly, a catalyzing agent with properties suggested above does exist. It is a subatomic particle commonly called the "muon" and represented by the symbol μ. This particle appears as a product in various types of high energy nuclear reactions and possesses the following properties:

> charge: $q_\mu = -1.6 \times 10^{-19}$ C   (same charge as an electron)
> mass: $m_\mu = 207 \, m_e$ (207 times as heavy as an electron)
> lifetime: $\tau_\mu = 2.2 \times 10^{-6}$ s   (it is unstable) .

The property that a muon has the charge of an electron but a much larger mass means that it can enter into an orbit of a hydrogen atom with a Bohr radius 207 times smaller than that for an electron; we suggest this comparison between a conventional hydrogen atom and a muonic hydrogen atom in Fig.7.7.

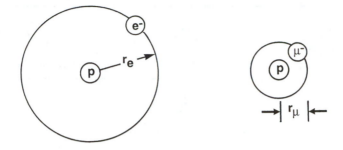

Fig. 7.7: A conventional hydrogen atom (left) and a muonic hydrogen atom (right); r_e = 207 r_μ and therefore this drawing is not to scale.

A consequence of a much tighter muon orbit around a hydrogen nucleus is that to another hydrogen ion or hydrogen atom, the muonic hydrogen atom appears like an oversized and overweight neutron. Hence, this "oversize" neutron might approach another hydrogen ion or atom more closely because of a reduction of the repulsive Coulomb forces. Then, when this conventional hydrogen and the muonic hydrogen are sufficiently close to "notice" the details of spatial charge variation, they are already close enough for nuclear forces of attraction to dominate and to bring about a fusion event.

Several additional points are important in this context of muon catalyzed d-t fusion. Reactions of the type depicted in Fig.7.8 have been experimentally confirmed in liquid hydrogen at temperatures in the 300 K to 900 K range; the implication therefore is that muon catalyzed fusion may be sustained in a temperature environment more like that of existing fission reactors. Then, an appropriate accelerator for muon production has to be associated close to the muon-fusion chamber suggesting a system configuration similar to that of an ion beam-sustained inertial confinement scheme. Finally, the muon mean life-time of 2.2 μs is unaffected by whether it is bound to a nucleus or not and hence demands that the various muonic induced reactions generally proceed at a fast rate.

Problems

7.1   Use $<\sigma v>_{dt}$ from Appendix C for 10 keV and 100 keV and calculate R_{fu} for a 50:50% and 25:75% mixture of deuterium and tritium for which $N_d + N_t = N = 10^{21}$ m^{-3}.

7.2   With the branching of the d-d fusion reactions occurring with essentially equal probability, determine the deuterium destruction rate and the helium-3 production rate.

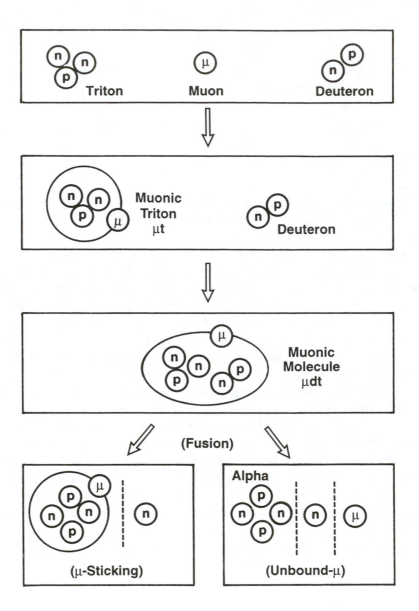

Fig. 7.8: Schematic of a muon catalyzed d-t reaction sequence.

7.3   For d-d catalyzed fusion, what are the ratios of (a) N_h / N_d, (b) N_t / N_d, (c)R_{+n} / $R_{+\alpha}$, (d) R_{+p} / $R_{+\alpha}$, and (e) $R_{+\alpha(d-t)}$ / $R_{+\alpha(d-d-p)}$.

7.4   Formulate reaction rate expressions for d, t and h as suggested in Table 7.2.

7.5   Undertake an analysis similar to that leading to the results of Fig.7.3 for the case of F = 0.

7.6   Fission reactors possess power densities of about 10^7 Wm^{-3}. For this power range, determine the required particle density for d-t fusion at 10 keV.

7.7   Consider a catalyzed deuterium reactor operating under the conditions $N_i=N_e=10^{20}$ m^{-3}, kT=10 keV. Calculate the equilibrium concentrations of all three species (d, t, and h).

7.8   Physics requirements for d-h fusion are clearly more demanding than for d-t. Some advantages often cited involve environmental, safety, and cost factors. Discuss how these factors are advantageous and whether or not the advantages outweigh the disadvantages. Do you foresee any advances in physics that might significantly affect any trade-offs? Discuss which confinement system might be best suited for burning d-h.

7.9   It has been suggested that running a "lean" deuterium mixture (e.g. 30 % d, 70 % h) could further reduce neutron production. However, this is a trade-off against power density. Suggest, using equations and sketches of graphs, how to determine an "optimum" mixture ratio. (You may assume that a 1000 MW-electric plant is desired and that this power level is fixed; describe how to find what mixture ratio would provide 1000 MW$_e$ for the least input power.)

PART III ENERGETICS, CONCEPTS, SYSTEMS

8.   Fusion Reactor Energetics

A fusion energy system will contain energy in various forms. The relative magnitude of each form and an understanding of how changes in each energy component can be affected need to be considered and assessed. We begin with an overall system energy balance and subsequently examine selected details of energy forms and transformations.

8.1   System Energy Balance

A fundamental requirement of any power system is that it be a net energy producer. That is, during some time interval τ, we require the overall net energy

$$E^*_{net} = E^*_{out} - E^*_{in} > 0 \tag{8.1}$$

where E^*_{in} is the total energy externally supplied to sustain fusion reactions and associated processes and E^*_{out} is the total recovered energy. Note that the asterisk notation is again used to indicate magnitudes referring to the entire reaction volume as opposed to the respective density expressions. While energy can exist in various forms–thermal, kinetic, radiation, electrical, magnetic, etc.–common units need to be employed in energy balance equations; for that, energy in electrical form is particularly useful.

   In Fig.8.1, we depict a power plant's energy flow in a simple general form. The energy eventually delivered to the reaction chamber differs from the provided input energy, E^*_{in}, by a factor η_{in} which, obviously, is the efficiency of converting the different forms of energy involved and of coupling specific energy forms into the fusion plasma. Analogously, the energy released from the fusion domain, which may consist of thermal energy as well as electromagnetic and neutron radiation, will have to be converted–for external use–into electricity with some overall efficiency η_{out} typical for the employed energy conversion cycles.

   Recall the connection between energy and power $P_0^* = dE_0^*/dt$. Indeed, since time variations are of dominant interest, particularly for pulsed systems, it is common in energy accounting to consider the time variations of power. Thus, the basic energy balance Eq.(8.1), may equally be written

$$\int_0^\tau \left(\frac{dE^*}{dt} \right)_{net} dt = \int_0^\tau \left(\frac{dE^*}{dt} \right)_{out} dt - \int_0^\tau \left(\frac{dE^*}{dt} \right)_{in} dt > 0 . \tag{8.2}$$

   We proceed now to identify individual power or energy components

123

generally associated with a fusion reactor plasma.

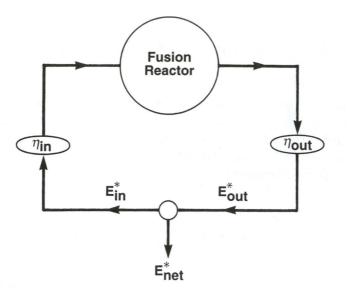

Fig. 8.1: Depiction of energy components associated with a general fusion power plant.

It is useful to recognize that, unlike fission reactors which sustain a fission chain by *particle* feedback (neutrons) requiring only a critical mass, fusion reactors will need a critical plasma energy concentration. For that, energy has to be initially supplied to the plasma which, upon reaching the critical heat condition, will be capable of sustaining the fusion processes by self-heating due to fusion power deposition in the plasma. Note therefore, that the nuclear fusion chain is maintained by *energy* feedback.

The achievement of the mentioned critical energy concentration in the plasma is yet an ongoing task in fusion research. Depending on the fusion device and fuel cycle deployed, today's fusion experiments are far from demonstrating energy self-sufficiency.

In general, the fusion plasma energy will be balanced by external (auxiliary) and internal (fusion power deposition) heating against the several losses due to radiation and other leakage. Considering the thermal energy content in the total plasma volume, that is we take here $E^* = E_{th}^*$, we integrate its temporal variation rate over a typical fusion burn time τ_b to relate

$$\int_0^{\tau_b} \frac{dE_{th}^*}{dt} dt = E_{aux}^* + E_{fu}^* - E_n^* - E_{rad}^* - \int_0^{\tau_b} \frac{E_{th}^*}{\tau_{E^*}} dt \qquad (8.3)$$

where E_{aux}^* is the auxiliary heat supply, E_{fu}^* represents the fusion energy production, E_n^* is the energy of fusion neutrons escaping from the plasma, and

E_{rad}^* denotes the radiation losses–all over time τ_b; the last term of Eq.(8.3) accounts for energy leakage through plasma transport processes as suggested by Eq.(6.53). Inspection of Fig.8.1 suggests taking

$$E_{aux}^* = \eta_{in} \, E_{in}^* \tag{8.4}$$

with η_{in} denoting the efficiency of input energy coupling to the plasma. Further, since the ratio

$$\frac{E_n^*}{E_{fu}^*} = 1 - f_c \tag{8.5a}$$

is a constant for the fusion cycle considered (see Table 7.1), we take

$$E_{fu}^* - E_n^* = f_c \, E_{fu}^* \tag{8.5b}$$

with f_c representing the charged particle fraction of fusion product energy (recall that $f_{c,dt} = 0.2$ only). Commonly, the definition of the plasma fusion multiplication or, briefly called the plasma Q-value,

$$Q_p = \frac{E_{fu}^*}{\eta_{in} \, E_{in}^*} \tag{8.6}$$

is introduced in Eq. (8.3) which then reads

$$\int_0^{\tau_b} \frac{dE_{th}^*}{dt} \, dt = \left(f_c + \frac{1}{Q_p} \right) E_{fu}^* - E_{rad}^* - \int_0^{\tau_b} \frac{E_{th}^*}{\tau_E^*} \, dt \, . \tag{8.7}$$

Note that Q_p is a measure for how efficiently an energy input to the plasma is converted into fusion energy. In the case of an energetically self-sufficient fusion plasma–later on we will call this state an ignited one–that is when the plasma meets its power demands by the fusion energy deposition alone and hence auxiliary heating is no longer required, the plasma Q-value approaches infinity as $E_{in}^* \rightarrow 0$. The left hand side of Eq. (8.7) can be immediately integrated to give

$$\int_0^{\tau_b} dE_{th}^* = E_{th}^*(\tau_b) - E_{th}^*(0) \, . \tag{8.8}$$

Obviously then, if the plasma is restored to the same energetic state after the burn, that is if $E_{th}^*(\tau_b) = E_{th}^*(0)$, the fusion energy generated in this burn is found from Eq.(8.7) to balance the losses by

$$E_{fu}^* = \frac{E_{rad}^* + \int_0^{\tau_b} \frac{E_{th}^*}{\tau_E^*} \, dt}{f_c + \frac{1}{Q_p}} \, . \tag{8.9}$$

For the case of energy self-sufficiency, $Q_p \rightarrow \infty$, the fusion energy delivered to the plasma via the charged reaction products is seen to balance the total energy loss from the plasma. Considering a d-t plasma and collecting the relevant power densities previously derived, we use the explicit expressions given in Eqs.(2.21a),

(7.2b), (3.44) and (5.88), whereby we suggest E_{rad}^* to be due to bremsstrahlung and cyclotron radiation, to rewrite Eq.(8.9) for $Q_p \rightarrow \infty$ yielding

$$f_{c,dt} \int_V d^3r \int_0^{\tau_b} R_{dt}(\mathbf{r},t) Q_{dt}\, dt = \int_V d^3r \left[\int_0^{\tau_b} \left(P_{br} + P_{cyc}^{net} \right) dt + \int_0^{\tau_b} \frac{E_{th}(\mathbf{r},t)}{\tau_E(\mathbf{r},t)} dt \right] \qquad (8.10)$$

where, in analogy to Eq. (7.19), the global energy loss has been replaced by the volume-integrated local leakage according to

$$\int_V d^3r \frac{E_{th}(\mathbf{r},t)}{\tau_E(\mathbf{r},t)} = \frac{E_{th}^*(t)}{\tau_{E^*}(t)} \qquad (8.11)$$

with $\tau_E(\mathbf{r},t)$ representing the local energy confinement time.

As mentioned previously, the plasma state described by Eq.(8.10), that is when the internal fusion power deposition is capable of balancing the plasma power losses, is commonly referred to as ignition. An auxiliary energy supply is thus no longer needed.

For a fusion reactor operating in steady state the time integrals in Eq.(8.10) may be omitted thus leaving behind a global plasma power balance. Further, if ignition is assumed to occur everywhere throughout the plasma volume (e.g. in a homogeneous plasma), then we may disregard the volume integration as well and finally derive the local d-t fusion ignition condition

$$f_{c,dt} P_{dt}(N_i, T_i) = P_{br}(N_i, N_e, T_e) + P_{cyc}^{net}(N_e, T_e) + \frac{3}{2} \frac{(N_i kT_i + N_e kT_e)}{\tau_E} \qquad (8.12)$$

where the thermal plasma energy density has been taken as

$$E_{th,j} = \tfrac{3}{2} N_j kT_j, \quad j = i, e \qquad (8.13)$$

according to the average particle energy in a Maxwellian distribution, Eq.(2.19c), multiplied by the respective number density. Explicit expressions for the several power terms in Eq.(8.12) may be readily introduced via their definitions in previous sections and thus reveal the complex interrelation between the plasma density and its temperature as required for ignition.

8.2   Plasma Heating

While the discussion in Sec. 4.6 and the confinement requirement suggested there, Eq.(4.12), provide some essential conceptual information about magnetically confined fusion, a sufficiently high plasma temperature has to be attained for sufficient fusion reactions to occur, Sec. 2.5. To attain this state, a neutral gas is heated and thereby ionized, as the gas kinetic temperature surmounts the respective ionization energy potentials. Upon this plasma formation, the heating process is still to be continued in order that the plasma approach the 10 keV temperature range favourable for a reasonable <σv>. When these plasma temperatures are reached, their sustainment has to be considered. Of the three processes mentioned, only ionization is elementary while both heating and high temperature sustainment are far more difficult.

The initial phase of a burn cycle, that is the attainment of sufficiently high temperature, as well as subsequently the sustainment of those conditions, can be accomplished by means such as the following:

(a) Resistive heating: this process involves Ohmic heating due to an electric current in the plasma.

(b) Compression: mechanical and/or magnetic forces are used to compress the plasma adiabatically and thus raise its temperature.

(c) Electromagnetic wave heating: electromagnetic waves from lasers or radiofrequency generators may be used to deposit energy in the plasma.

(d) Beam injection: neutral particles or pellets are injected and deposit their energy by collisional effects.

(e) Internal heating: charged fusion products collisionally transfer most of their birth energy to the plasma ions and electrons.

Resistive heating is based on Ohmic energy dissipation effects. The power deposited in a unit volume of a plasma by this method is given by

$$P_{res} = \eta I^2 \qquad (8.14)$$

where I is the current density; the parameter η is the plasma resistivity and possesses a dependence on a number of collisional effects and is of the form

$$\eta \propto kT^{-3/2} . \qquad (8.15)$$

The feature that the resistivity of a plasma decreases with increasing temperature means that Ohmic heating becomes progressively less effective at higher temperatures. Above about 1 keV, supplementary heating must be employed unless—as is possible for some system concepts—massive currents are applicable.

Compression methods of heating can be classified into two broad categories. If it occurs very rapidly—on the scale of 10^{-6} s or less—then it is an implosion and complex gas dynamics and shock wave considerations need to be introduced. If the compression occurs over longer intervals compared to the speed of thermal energy transfer, but still short relative to radiation losses, then the compression is adiabatic and the necessary relation

$$pV^\gamma = \text{constant} \qquad (8.16)$$

holds. Here γ is the relevant adiabatic gas coefficient, as in Eq.(6.26).

Since a plasma consists of an aggregation of numerous moving electrical charges, it is subject to collective interactions exhibiting typical resonance effects. Therefore, coupling of high power (high frequency) waves to the plasma appears to be an effective heating mechanism. A most favourable application of such is the electromagnetic coupling via waves having the ion cyclotron frequency $\omega_{g,i}$, Eq. (5.12). The ion motion can thus be resonantly enhanced to high kinetic energies. The irradiation by those high frequency waves is usually performed at the frequency $\omega_{g,i}$, or the harmonics $2\omega_{g,i}$, or $3\omega_{g,i}$ (30-100 MHz depending on the magnetic field strength). The absorption of the irradiated electromagnetic energy increases with higher ion temperature T_i and can be managed most effectively by minimizing the distance between the wave antenna and the plasma edge.

Aside from this so-called ion cyclotron resonance heating (ICRH), there is, of course, also the possibility of electron cyclotron resonance heating (ECRH) which will require frequencies in the range of 28-140 GHz. Note that the specific heating of a particular plasma species may lead to a significant difference in T_e and T_i; it is the latter which is to be raised for a sufficient fusion reactivity.

Beam injection, particularly neutral atoms or macro particles, has proven to be an important and effective means of heating. The mechanism of energy transfer in the plasma initially involves the conversion of the high energy neutrals into high energy ions by charge exchange and impact ionization; the resultant fast ions subsequently transfer part of their energy to the plasma ions and electrons by Coulomb collisions.

The rate at which injected fast ions transfer their kinetic energy E_f to the plasma involves two distinct components. The average energy transfer rate to electrons is approximately of the form

$$< \frac{dE}{dt} >_{f \to e} \approx A_e \, N_e \, T_e^{-3/2} \, E_f \ , \tag{8.17}$$

while the transfer rate to ions may be approximated by

$$< \frac{dE}{dt} >_{f \to i} \approx A_i \, N_i \, E_f^{-1/2} \tag{8.18}$$

with both A_e and A_i constants for the respective species and specified beam particles. Thus, at a high ion beam energy, most of the energy transfer is to electrons whereas at lower energies the thermal ions get a larger share.

We note that fusion product heating will occur due to the same collisional effect, i.e. Coulomb scattering. Hence the same energy transfer rates, as given by Eqs.(8.17) and (8.18), apply to the slowing down of the charged fast fusion products in the plasma which in turn is heated. In d-t fusion, this fusion power deposition is called alpha-heating, attributable to the charged particles released from the nuclear reaction. Very high energetic fusion products, e.g. 15 MeV protons from d-^{3}He fusions, can as well lose energy by nuclear elastic scattering, which will result in discrete energy transfers to the plasma ions.

8.3   Lawson Criterion

It is often desirable to use easily understood and readily recognizable parameters as indicators of the merit of a particular system or process. A widely used formulation that provides a test of the energy balance for fusion devices is known as the Lawson Criterion. It is an algebraic relationship based on the assertion that the recoverable energy from a fusion reactor must exceed the energy which is supplied to sustain the fusion reaction; that is, $E_{out}^* > E_{in}^*$ must hold during a representative time interval τ which may be the cycling time for a pulsed fusion device or a typical period of time if steady-state operation is envisaged.

We may associate the recoverable energy from a fusion device with three forms of energy: the fusion reaction energy release, E_{fu}^*, all radiation losses, E_{rad}^*, and the thermal motion of the particles, E_{th}^*, Fig. 8.2. Since no energy conversion can take place with 100% efficiency, the fraction of the energy that is recovered can be written as

$$E_{out}^* = \eta_{fu} E_{fu}^* + \eta_{rad} E_{rad}^* + \eta_{th} E_{th}^* \tag{8.19}$$

where the η's are the respective conversion efficiencies for each energy component.

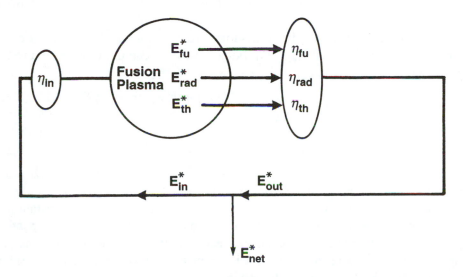

Fig. 8.2: Energy flow for an arbitrary fusion reactor.

The energy supplied to the device in order to sustain the fusion reactions, E_{in}^* in Fig.8.2, appears as the thermal motion of the particles and as radiation:

$$\eta_{in} E_{in}^* = E_{rad}^* + E_{th}^* . \tag{8.20}$$

Here η_{in} is the fraction of E_{in}^* which is coupled to the plasma and amounts to E_{th}^* and E_{rad}^*; that is, only the fraction $\eta_{in} E_{in}^*$ is actually deposited to heat and sustain the fuel at a constant temperature.

The energy viability statement, Eq.(8.1), can now be written as

$$\eta_{fu} E_{fu}^* + \eta_{rad} E_{rad}^* + \eta_{th} E_{th}^* > \frac{E_{rad}^* + E_{th}^*}{\eta_{in}} \tag{8.21}$$

or

$$\eta_{in}\,\eta_{out}\left(E^*_{fu} + E^*_{rad} + E^*_{th}\right) > E^*_{rad} + E^*_{th} \tag{8.22}$$

where, for reasons of algebraic convenience, we have taken η_{out} as an average conversion efficiency

$$\eta_{out} = \frac{\sum\limits_{\ell}\eta_{\ell}\,E_{\ell}}{\sum\limits_{\ell}E_{\ell}}\,,\qquad \ell = fu,\,rad,\,th\,. \tag{8.23}$$

Consider now the idealization that the system operates in a mode characterized by a constant power production during a global energy confinement time, τ_{E^*} such that the global energy terms, E^*_{ℓ}, can be readily evaluated from the respective power densities by

$$E^*_{\ell} = \tau_{E^*}\int_V P_{\ell}(\mathbf{r})\,d^3r\,. \tag{8.24}$$

Also, for present purposes it is assumed that radiation losses are due to bremsstrahlung only so that $P_{rad} = P_{br}$. Equation (8.22) can therefore be expressed in terms of these steady power flows for time τ_{E^*} to give

$$\eta_{in}\,\eta_{out}\int_V d^3r\left(\tau_{E^*}P_{fu} + \tau_{E^*}P_{br} + 3NT\right) > \int_V d^3r\left(\tau_{E^*}P_{br} + 3NT\right) \tag{8.25}$$

where we have substituted for the thermal energy density according to Eq. (8.13),

$$E_{th}(\mathbf{r}) = \tfrac{3}{2}\left(N_iT_i + N_eT_e\right) = 3NT \tag{8.26}$$

assuming $N_i = N_e = N$ and $T_i \approx T_e = T$. Further, we have suppressed Boltzmann's constant k in our notation, since from now on we will always refer to T as kinetic temperature in units of eV or keV and hence already incorporate the constant k.

Next we introduce in Eq. (8.25) the density and temperature dependent expressions for P_{fu} and P_{br} discussed in preceding chapters and–for reasons of a simplified demonstration of the parameter interrelation–we assume homogeneity throughout the plasma volume V to obtain

$$\eta_{in}\,\eta_{out}\left(\frac{N_aN_b}{1+\delta_{ab}}<\sigma v>_{ab}Q_{ab}\,\tau_{E^*} + A_{br}N^2\sqrt{T}\,\tau_{E^*} + 3NT\right) > A_{br}N^2\sqrt{T}\,\tau_{E^*} + 3NT \tag{8.27}$$

with the Kronecker-δ introduced in the fusion rate density in order to account for the case of indistinguishable reactants, i.e. when a = b. Here $\eta_{(\,)}$, Q_{ab} and A_{br} are constants, whereas $<\sigma v>_{ab}$ is a function of temperature (see Fig.7.5). Taking a 50:50% fuel mixture, i.e., $N_a = N_b = N/2$, and solving the above inequality for the product $N\tau_{E^*}$ yields the well-known Lawson Criterion for energy viability

$$N\tau_{E^*} > \frac{3(1-\eta_{in}\,\eta_{out})T}{\eta_{in}\,\eta_{out}\dfrac{<\sigma v>_{ab}(T)Q_{ab}}{4(1+\delta_{ab})} - (1-\eta_{in}\,\eta_{out})A_{br}\sqrt{T}} \tag{8.27}$$

representing a rough, but useful reactor criterion.

Thus, energy viability requires that the density N times the global energy confinement time, τ_{E*}, must exceed a particular function of ion temperature. With all constants known and for illustrative purposes taking $\eta_{in} \cdot \eta_{out} \approx 1/3$, as was suggested by Lawson, it becomes possible to determine this lower bound of $N\tau_{E*}$ as a function of the plasma temperature as shown in Fig.8.3.

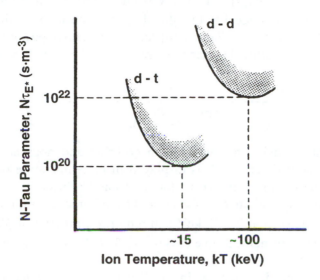

Fig. 8.3: Lawson criterion bounds for d-t and d-d fusion.

As indicated in Fig.8.3, the lower bound of $N\tau_{E*}$ attains a minimum value at a specific temperature. This is a useful result and defines the combination of particle densities, confinement times, and temperature needed for reaction energy break-even conditions. The extremum point in the case of d-t fusion, $N\tau_{E*} \sim 10^{20}$ m^{-3}s at approximately 15 keV was commonly cited as an initial experimental objective. Note also that no particular fusion design was necessary in the derivation of this criterion.

Equation (8.28) does not contain all relevant processes. For example, cyclotron radiation emission is not included and it may be difficult to sustain the various operational parameters as constants during the time interval τ_{E*}. Nevertheless, Eq.(8.28) is a useful and widely employed criterion. For commercial power applications, it would be necessary to exceed the minimum Lawson limit by perhaps a factor of ten or better.

8.4 Ignition and Break-Even

While the Lawson Criterion represents a reactor criterion, i.e. it refers to the energy viability of the entire plant, we can as well derive a criterion for the energy viability of the fusion plasma. The latter may be deemed as energetically viable when external power has no longer to be delivered to it. This so-called ignited plasma state can be shown for a homogeneous plasma, as was also assumed for Eq.(8.28), with help of Eqs.(8.10) and (8.11) to be characterized by

$$f_{c,dt} P_{dt} \tau_{E^*} \geq \left(P_{br} + P_{cyc}^{net} \right) \tau_{E^*} + 3NT \qquad (8.29)$$

where again $N_i = N_e = N$, $T_i \approx T_e = T$ was taken and the time integration was performed over a period equal to the global energy confinement time τ_{E^*}. As in deriving Eq.(8.28), we can similarly substitute for the above power terms by their explicit density-temperature dependent expressions, take $N_d = N_t = N/2$ and solve for $N\tau_{E^*}$ to obtain

$$\left(N\tau_{E^*} \right)_{dt} \geq \frac{3T}{f_{c,dt} Q_{dt} \dfrac{<\sigma v>_{dt}(T)}{4} - A_{br}\sqrt{T} - \dfrac{A_{cyc} B^2 \psi T}{N}} \qquad (8.30)$$

which represents a fusion plasma ignition criterion and does not contain any energy conversion efficiencies. It is displayed in its temperature dependence in Fig. 8.4. A contour plot similar to the Lawson Criterion becomes evident, however featuring quantitative differences. The minimum ignition temperature for $N = 10^{20}$ m^{-3} is seen at $T \approx 30$ keV and requires a product of plasma density and global energy confinement time of $N\tau_{E^*} \approx 2.7 \times 10^{20}$ m^{-3}s and hence $\tau_{E^*} \approx 2.7$ seconds.

We note that the denominator in Eq.(8.30) can become negative for high temperatures where the cyclotron radiation terms takes over to dominate the plasma energetics. Such a regime is associated with a negative plasma energy balance and can therefore not be ignited. For that temperature range, Eq. (8.30) has no meaning. The more stringent ignition requirements with increasing T are evident from Fig. 8.4 where the ignition contours tend towards infinity as the plasma temperature approaches the critical value T_{crit}, as indicated for the case of $N = 10^{20}$ m^{-3}.

Another definition often used for classifying the fusion reactor operation is the scientific (energy) break-even which means that the total fusion energy production amounts to a magnitude equal to the effective plasma energy input, i.e.

$$\frac{E_{fu}^*}{\eta_{in} E_{in}^*} = Q_p = 1 . \qquad (8.31)$$

This break-even condition can be analogously derived from Eq. (8.9) for $Q_p=1$ and is also demonstrated in Fig. 8.4. Recall that the ignition condition is associated with $Q_p \rightarrow \infty$.

The performance of a high $N\tau_{E^*}$ is not the only goal in fusion reactor research; as the preceding discussion suggests, it is additionally required for a viable fusion plasma regime that, simultaneously with $N\tau_{E*}$, the plasma temperature is established at a sufficiently high level. Hence, presently, the so-called triple product $TN\tau_{E*}$ is used most often for qualifying recent achievements of fusion experiments. This last ignition criterion is easily obtained by just multiplying Eq.(8.30) with T on both sides. Expectedly, it does not introduce a new characteristic, but rather exhibits a similar temperature dependence as in Fig. 8.4.

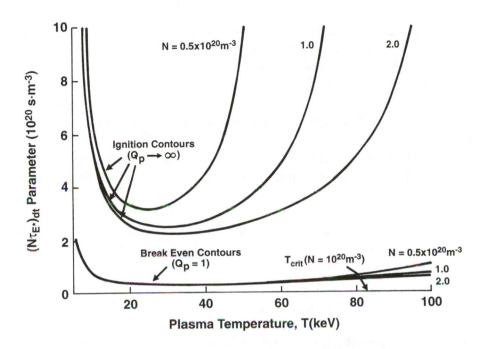

Fig. 8.4: Ignition and energy break-even conditions for d-t fusion as a function of plasma temperature. Since P_{cyc}^{net} was also accounted for, in which $\psi = 0.01$ has been taken, a density dependence of $N\tau_{E*}$ is observed.

Problems

8.1   Reformulate the Lawson Criterion which incorporates the direct conversion of the alpha particle into electricity with 95% efficiency.

8.2   Discuss and compare the plasma energy $kT \sim 15$ keV associated with the minimum $N\tau_{E*}$ requirement given by the Lawson Criterion, with the recognition that a kinetic energy of $E_k \sim 200$ keV is required by hydrogen ions to overcome their Coulomb repulsion in order to fuse.

8.3   Develop Lawson Criteria for a d-d plasma (include both branches of the d-d reaction), and plot $N\tau$ vs T. Assume a fusion power to electrical conversion efficiency of 30%.

8.4   Develop Ignition Criteria for a d-t, d-h and d-d plasma, and plot $N\tau$ vs T for each. Assume alpha-particle heating of the d-t plasma, alpha-particle and proton heating of the d-h plasma, and triton, proton and h heating of the d-d plasma.

8.5   Derive–analogously to Eq. (8.28)–a more realistic MCF reactor criterion accounting also for cyclotron radiation losses Display graphically its temperature dependence, $N_e\tau_{E^*}$ (T) for the two cases
    (a) d-t: $N_d = N_t = N_e/2$, $T_i \approx T_e$, $\psi/N_e = 10^{-23}$ m^3, B = 5 T, $\eta_{in} = 0.5$ and η_{out}=0.35;
    (b) d-h: $N_d = N_h = N_i/2$, $T_i \approx T_e$, $\psi/N_e = 10^{-22}$ m^3, B = 10 T, $\eta_{in} = 0.5$ and η_{out}=0.7.
Note: For the derivation of the criterion, use fractions of the ion density such that $N_j = \kappa_j N_i$ with j denoting the considered ion species. You should finally obtain:

$$\left(N_e\tau_E^*\right)_{ab} > \frac{\dfrac{3}{2}\left(\dfrac{T_i}{\Sigma\kappa_j Z_j} + T_e\right)}{\dfrac{\eta_{in}\eta_{out}}{1-\eta_{in}\eta_{out}}\left[\dfrac{\kappa_a\kappa_b}{\Sigma\kappa_j Z_j\left(1+\delta_{ab}\right)}\dfrac{<\sigma v>_{ab}}{} Q_{ab} - \dfrac{\Sigma\kappa_j Z_j^2}{\Sigma\kappa_j Z_j}A_{br}\sqrt{T_e} - A_{cyc}B^2\dfrac{\psi}{N_e}T_e\right]},$$

for j = a, b, impurities. Compare the plots from (a) and (b) with those that result if cyclotron radiation losses are neglected.

8.6   Discuss the physical differences between (i) the Lawson reactor criterion, (ii) the ignition criterion and (iii) the break-even condition. What fraction of the entire plasma loss power can be made up for by α-particle heating in steady state operation, when $Q_p = 5$?

9. Open Magnetic Confinement

The concept and development of fusion reactors based on the confinement of a high temperature plasma with magnetic fields has been approached from many perspectives. However, utilizing the effects of magnetic fields on charged particles has, in general, led to systems wherein the magnetic field lines are either open–as in the case of the magnetic mirror previously discussed–or closed–as for the case of toroidal devices. We first consider the former, following historically the development of magnetic confinement devices which began with open field line systems.

9.1 Magnetic and Kinetic Pressure

In Chs. 4 and 5, we have introduced some basic considerations of magnetic confinement. To begin this more detailed look, we consider the containment capacity of magnetic fields and subsequently discuss specific reductions.

The macroscopic fluid description of a plasma does not clearly suggest how individual charged particles could be confined by a magnetic field. Even if the external **B**-field is taken to consist of smooth straight field lines, the plasma could generate internal fields which might cause drift motion leading to particle escape from the confinement region. An effective containment will require the plasma particles to be confined to some region for a sufficient time period. We may therefore assert that, in general, this is associated with a state of the macroscopic fluid in which all identified forces are exactly in balance providing thereby a time-independent solution. We consider therefore Eq.(6.25) for steady state, i.e. $\partial \mathbf{V}_j/\partial t = 0$, and obtain for the isotropic case $(\mathbf{V}_j \cdot \nabla)\mathbf{V}_j = 0$ in the absence of an electric field, the defining equation

$$\nabla p_j = \rho_j^c \mathbf{V}_j \times \mathbf{B}. \tag{9.1}$$

Imposing this equation for plasma ions, $j = i$, and electrons, $j = e$, we write for the total plasma pressure gradient

$$\nabla p = \nabla(p_i + p_e)$$
$$= (\rho_i^c \mathbf{V}_i + \rho_e^c \mathbf{V}_e) \times \mathbf{B} \tag{9.2}$$
$$= \mathbf{j} \times \mathbf{B}.$$

Evidently, $\mathbf{j}$ is the electric current density which must also satisfy Maxwell's equation

$$\mathbf{j} = \nabla \times \frac{\mathbf{B}}{\mu_o} \ . \tag{9.3}$$

We note that Eq.(9.2) asserts that in equilibrium, there exists an equivalence between the pressure gradient force and the Lorentz force. Further, the electric current density $\mathbf{j}$ and the magnetic field $\mathbf{B}$ are perpendicular to ∇p; that is, as suggested in Fig. 9.1, $\mathbf{j}$ and $\mathbf{B}$ lie on an isobaric surface because everywhere ∇p is a normal to the surfaces p = constant.

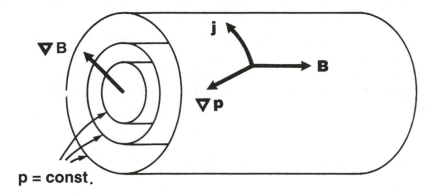

Fig. 9.1: Depiction of an electric current and magnetic field–at plasma equilibrium–on surfaces of constant pressure.

Introducing Eq.(9.3) into (9.2) leads, upon some differential vector analysis, to the expression

$$\nabla p = \frac{1}{\mu_o}(\mathbf{B} \cdot \nabla)\mathbf{B} - \frac{1}{2\mu_o}\nabla B^2 \tag{9.4}$$

which, for the case of straight, parallel field lines, reduces to

$$\nabla\left(p + \frac{B^2}{2\mu_o}\right) = 0 \tag{9.5a}$$

or, equivalently,

$$p + \frac{B^2}{2\mu_o} = \text{constant} . \tag{9.5b}$$

This relationship is an important characterization of a hydrodynamic equilibrium state. It makes evident that the total sum of kinetic pressure[*] and magnetic field energy density

$$\frac{E^*_{mag}}{V} = \frac{BH}{2} = \frac{B^2}{2\mu_o} , \tag{9.6}$$

---

[*] Note that from here on, the subscript *kin*, which was used in Eq. (4.11) for instruction and distinction purposes, is dropped.

where H is the local field strength, will be a constant everywhere within the confined plasma. Hence, the pressure gradient ∇p induces a current in the direction shown in Fig. 9.1 and which, by induction, causes an according decrease in the magnetic field in the plasma. From a particle perspective, we could as well have used Eq.(5.36) to derive a drift due to the pressure gradient force $-\nabla p$ with the velocity

$$\mathbf{v}_{D,\nabla p} = -\frac{\nabla p \times \mathbf{B}}{Nq\,B^2} \tag{9.7}$$

for a fluid element containing N particles. In view of the differences in ion and electron drifts, we define the electric current density by

$$\mathbf{j} = N_i q_i \mathbf{v}_{D,i} + N_e q_e \mathbf{v}_{D,e} = \frac{\mathbf{B} \times \nabla p}{B^2} \tag{9.8}$$

and label it the diamagnetic current. The above expression is essentially equivalent with the previous current density definition in Eq.(9.2) which, when cross-multiplied by $\mathbf{B}$, results in Eq. (9.8), recognizing that $\mathbf{j}$ is perpendicular to $\mathbf{B}$.

Characterizing the thermodynamic state of the plasma by the Ideal Gas Law

$$pV = N^* kT \tag{9.9}$$

with N^* as the total number of particles in volume V, we obtain an expression for the total kinetic pressure for an ensemble of ions and electrons as

$$p = \frac{N_i^*}{V} kT_i + \frac{N_e^*}{V} kT_e$$
$$= N_i kT_i + N_e kT_e \tag{9.10}$$
$$= (N_i + N_e)kT .$$

Both ions and electrons are, in this last expression, taken to be characterized by the same temperature $T_i = T_e = T$ for their Maxwellian distributions.

The extent of coupling between the magnetic pressure Eq.(9.6) and corresponding kinetic pressure Eq.(9.10) is a characteristic of a particular magnetic system and defines the dimensionless beta-parameter as the ratio of these two pressure terms:

$$\beta = \frac{(N_i + N_e)kT}{B^2 / 2\mu_o} . \tag{9.11}$$

In a plasma with a pressure gradient the magnetic field is, as suggested by Eq.(9.5b), low where the particle pressure–or mass density–is high and vice-versa. Some magnetic configurations exist which produce a diamagnetic current such that the internal magnetic field may, in a limited region, be reduced to a minimum value or even to $B \approx 0$; obviously, the local β-value would then approach infinity. As a consequence, it is common to define β as the ratio of maximum particle pressure–e.g. at the centre of the plasma cylinder, if we refer to Fig. 9.1–to the maximum magnetic pressure–e.g. at the outer surface of the plasma cylinder–thereby limiting the β-value to β ≤ 1. In most magnetic configurations, fusion plasma confinement requires an imposed magnetic pressure which significantly exceeds the intrinsic particle kinetic pressure. A low beta facility is typically characterized by β < 0.1. In Ch. 4 it was shown that the fusion power density varies as β^2 so that a high β facility is desirable recognizing however a limitation due to plasma instabilities.

9.2   Magnetic Flux Surfaces

Having established some characteristics and constraints with respect to particle pressure and magnetic pressure in a magnetic confinement fusion device, we now further examine some properties of the magnetic field itself. In Ch. 5, we discussed the motion of individual charged particles due to a variety of constant and spatially dependent magnetic fields. Here we investigate the case of a simple axially symmetric **B**-field, the consequent particle motions then recalled with some results from Ch. 5.

Specifically, we consider particle motion in a cylindrically symmetric configuration, i.e. $\partial/\partial\theta = 0$, with an axial magnetic field $\mathbf{B} = B_z\mathbf{k}$, as illustrated in Fig. 9.2. As suggested by electromagnetic theory, we take **B** to be represented by means of a vector potential **A**,

$$\mathbf{B} = \nabla \times \mathbf{A} .\tag{9.12}$$

This vector potential can be shown to be determined from the electric currents which generate **B** and is, in the stationary case, given by

$$\mathbf{A}(\mathbf{r}) = \frac{\mu_o}{4\pi} \int \frac{\mathbf{j}(\mathbf{r}')}{|\mathbf{r} - \mathbf{r}'|} d^3r'\tag{9.13}$$

where **r** is to be taken in cylindrical coordinates (r,θ,z). Since, in the geometry chosen here, the field generating currents point in the poloidal direction $\mathbf{e}_\theta$, it is obvious that **A** possesses only one non-zero component which is A_θ and thus yields

$$\mathbf{B} = \nabla \times A_\theta\mathbf{e}_\theta = -\frac{\partial A_\theta}{\partial z}\mathbf{e}_r + \frac{1}{r}\frac{\partial}{\partial r}(rA_\theta)\mathbf{k} .\tag{9.14}$$

In the absence of any other external field, the equation of particle motion is then, according to Eq.(5.1), described by

$$m\frac{d\mathbf{v}}{dt} = q\left\{\mathbf{E}_A + \left[\mathbf{v} \times \left(-\frac{\partial A_\theta}{\partial z}\mathbf{e}_r + \frac{1}{r}\frac{\partial}{\partial r}(r\,A_\theta)\mathbf{k}\right)\right]\right\}\tag{9.15}$$

with $\mathbf{E}_A$ representing that electrical field which is consistent with temporal variations of **B** and hence satisfies Maxwell's relation

$$\nabla \times \mathbf{E}_A = -\frac{\partial \mathbf{B}}{\partial t} = -\nabla \times \frac{\partial \mathbf{A}}{\partial t} \tag{9.16}$$

wherefrom

$$\mathbf{E}_A = -\frac{\partial \mathbf{A}}{\partial t} = -\dot{A}_\theta \, \mathbf{e}_\theta \tag{9.17}$$

is found. Here and in the following, the respective total time derivatives have been replaced by the dots-notation. Our interest here is in the θ-component of Eq.(9.15), which, using Eq. (9.17), is explicitly represented by

$$m(r\ddot{\theta} + 2\dot{r}\dot{\theta}) = -\frac{q}{r}\left(\dot{r}A_\theta + r\dot{A}_\theta + r\dot{r}\frac{\partial A_\theta}{\partial r} + r\dot{z}\frac{\partial A_\theta}{\partial z}\right) = -\frac{q}{r}\frac{d}{dt}\left(r(t)A_\theta(r,z,t)\right). \tag{9.18}$$

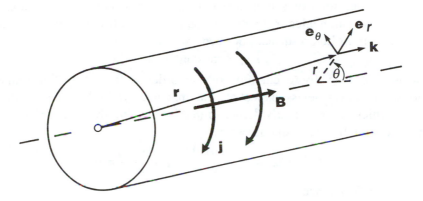

Fig. 9.2: Cylindrical geometry illustrating the relevant unit vectors for particle motion in an axial **B**-field generated by a poloidal current **j**.

Multiplication of the above equation by r leads to the angular momentum ℓ due to the rotational motion about the z-axis, i.e.

$$m\left(r^2\ddot{\theta} + 2r\dot{r}\dot{\theta}\right) + r\frac{q}{r}\frac{d}{dt}(rA_\theta) = \frac{d}{dt}\left(mr^2\dot{\theta} + qrA_\theta\right)$$

$$= \frac{d}{dt}(\ell) \tag{9.19}$$

$$= 0.$$

Evidently, ℓ is a constant of the motion and we write

$$rA_\theta\left(\frac{mr\dot{\theta}}{q \, A_\theta} + 1\right) = \frac{\ell}{q} \tag{9.20}$$

wherein A_θ needs to be determined from

$$B_z = (\nabla \times \mathbf{A})\cdot \mathbf{k} \equiv \frac{1}{r}\frac{\partial}{\partial r}(rA_\theta). \tag{9.21}$$

That is, specifically we have

$$A_\theta = \frac{1}{r} \int_0^r r' \, B_z(r') d'r' \; ; \tag{9.22a}$$

roughly assessing, A_θ can be approximated to the lowest order by

$$A_\theta \approx \tfrac{1}{2} r \, B_z(0) \tag{9.22b}$$

and inserted into Eq.(9.20) which, recognizing that the rotation is such that $sign(\dot{\theta}) = sign(q)$ and further that $r|\dot{\theta}| = v_\perp$ and here $mv_\perp/|q|B_z = r_g$, can then be written as

$$r \, A_\theta \left(1 + \frac{2 r_g}{r} \right) = \frac{\ell}{q} = \text{constant} . \tag{9.23}$$

Since the dimensions of a confined plasma are normally large compared to a gyroradius, we can neglect $r_g / r \ll 1$ in Eq.(9.23) and find that the trajectories of the particles must lie on surfaces defined by

$$r A_\theta = \text{constant} \tag{9.24}$$

allowing, however, for excursions of the order of $r_g \ll r$. Not immediately obvious, but nevertheless expected, the magnetic field lines lie within these surfaces which can be readily demonstrated by proving that the surface's normal is orthogonal to the field; that is, we get

$$\mathbf{B} \cdot \nabla(r A_\theta) = B_r \frac{\partial(r A_\theta)}{\partial r} + B_z \frac{\partial(r A_\theta)}{\partial z} = 0 \tag{9.25}$$

which is satisfied, since

$$B_r = \mathbf{e}_r \cdot (\nabla \times \mathbf{A}) = -\frac{\partial A_\theta}{\partial z} \tag{9.26a}$$

and

$$B_z = \mathbf{k} \cdot (\nabla \times \mathbf{A}) = \frac{1}{r} \frac{\partial}{\partial r}(r \, A_\theta) . \tag{9.26b}$$

These surfaces where $r A_\theta = $ constant may be labeled flux surfaces of the magnetic field and the particle's guiding centres move on them in the absence of other forces; this follows as a consequence of angular momentum conservation. Even if field gradients exist, which cause particle drifts across **B**-lines, the particles would, in a cylindrically symmetric plasma, remain on their flux surface since the underlying symmetry here constrains all gradients into the radial direction and, correspondingly, all drifts appear in poloidal directions. In this case, only collisions or anomalous transport processes can drive particles across the flux surfaces, which then is described by the perpendicular diffusion coefficient, Eqs.(6.18b) and (6.19). We further note that, according to Eq.(9.5b), the surfaces of constant B must also be surfaces of constant pressure and consequently, in most cases also of constant density and temperature.

9.3   Magnetic Mirror

A current-carrying solenoid provides the simplest way to establish a cylindrically homogeneous magnetic field to contain a plasma. While such a device is relatively easy to construct and to operate, it suffers from a serious deficiency: leakage of plasma particles through the open ends constitutes a significant loss of fuel ions and thermal energy. A reduction of this end-leakage can be accomplished by establishing an increasing magnetic field at the two ends so that many of the charged particles are trapped because of the imposed constraints on particle motion with regards to conservation of energy and the magnetic moment as discussed in Secs. 5.7 and 5.8. This concept resulted in the name "magnetic mirror" and is illustrated in Fig. 9.3.

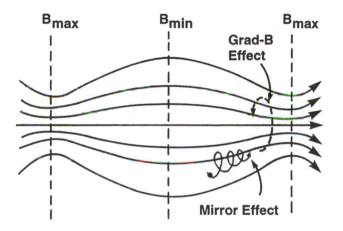

Fig. 9.3: Magnetic field lines in a magnetic mirror with an ion trajectory influenced by the mirror and grad-B effects also suggested.

Some important features of mirror fields can be described and analyzed on the basis of our preceding discussion of ion motion in homogeneous and inhomogeneous magnetic fields.

As discussed in Ch. 5, a positively charged ion spirals in a counterclockwise pattern about the direction of magnetic field lines; this is suggested in Fig. 9.3 for the case of a non-zero speed component of the ion parallel to the direction of the **B**-field. As the ion approaches the ends, it is increasingly subjected to drifts due to the inhomogeneity of the mirror field. However, note that these drifts occur in azimuthal directions and the particles are still bound to their magnetic surfaces discussed above. Thus they would be well confined in relation to the radial direction ($\perp$**B**), if not for an occurrent perpendicular diffusion due to collisional and/or turbulent transport, Eq.(6.19), resulting in particle migration out of the contained plasma region.

The more important particle losses, however, will appear in the direction parallel to **B**, that is through the ends of the solenoid. How effectively these losses can be reduced by squeezing the **B**-lines at the ends with a magnetic field strength increase, is discussed next.

The concept of trapping the ion in the magnetic "bottle" of a mirror field configuration can be formulated by drawing upon particle kinetic energy conservation and the constancy of the magnetic moment. As in Sec. 5.8, we write for an individual ion the energy in terms of the parallel and perpendicular velocity components, as established by the **B**-field lines, by

$$E_o = \tfrac{1}{2}mv_{\parallel}^2 + \tfrac{1}{2}mv_{\perp}^2 . \tag{9.27a}$$

With E_o a constant of motion and in the absence of other force effects, we can therefore depict the kinetic state of the ion anywhere in a collisionless plasma on a line in the $v_{\perp}^2 - v_{\parallel}^2$ plane, Fig.9.4a.

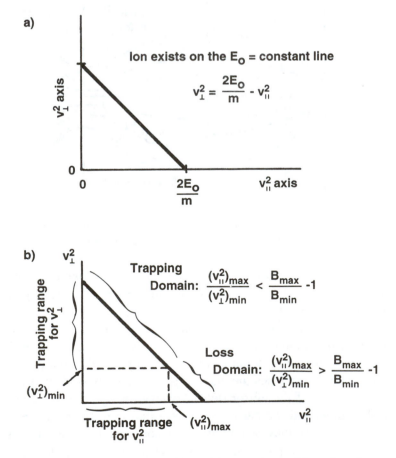

Fig. 9.4: Depiction of kinetic state of an ion and its trapping and loss domains on the $v_{\parallel}^2 - v_{\perp}^2$ plane.

Introducing the constant magnetic moment μ, Eq.(5.76), the energy conservation reads as

$$E_o = \tfrac{1}{2}mv_\parallel^2 + \mu B .$$ (9.27b)

Depending upon where in the mirror an ion exists, it will sense a different magnetic field–from a B_{min} in the mid-plane to a B_{max} in the mirror plane, Fig. 9.3. This range of magnetic field variation can be introduced into Eq.(9.27b) by recognizing that, with E_o a constant, $v_\parallel$ will be a maximum when B is a minimum,

$$E_o = \tfrac{1}{2}m\left(v_\parallel^2\right)_{max} + \mu B_{min}$$ (9.28)

and therefore $(v_\parallel)_{max}$ occurs at the central plane, Fig. 9.3. Conversely, $v_\parallel$ will be a minimum when B is a maximum. The condition for trapping of particles–that is stopping their parallel motion before or at the B_{max}-plane and thus not allowing them to escape through the ends is, in the absence of collisional effects, therefore

$$v_\parallel \Big|_{B \leq B_{max}} = 0 .$$ (9.29)

Using this relation in Eq.(9.27b) the trapping condition becomes

$$E_o \leq 0 + \mu B_{max}$$ (9.30)

and hence requires that

$$\tfrac{1}{2}m\left(v_\parallel^2\right)_{max} + \mu B_{min} \leq \mu B_{max}$$ (9.31)

wherein the equality sign yields the maximum $v_\parallel^2$ permissible for trapping. Rearranging Eq.(9.31) yields

$$\frac{\tfrac{1}{2}m\left(v_\parallel^2\right)_{max}}{\mu B_{min}} \leq \frac{B_{max}}{B_{min}} - 1$$ (9.32)

and using the definition of the magnetic moment provides the useful relation for the speed components at the mid-plane:

$$\frac{\left(v_\parallel^2\right)_{max}}{\left(v_\perp^2\right)_{min}} = \left(\frac{v_\parallel^2}{v_\perp^2}\right)_{mid\text{-}plane} \leq \frac{B_{max}}{B_{min}} - 1 .$$ (9.33)

This ratio identifies a domain on the E_o = constant line in Fig. 9.4b with which one may associate the ranges of $v_\parallel^2$ and $v_\perp^2$ which will ensure trapping of the ion. Note that this also conforms to an intuitive appreciation of ion trapping: the ion should have a relatively small $v_\parallel$-speed component so as to decrease its chances of escape through the ends. The probability of particle trapping or of particle loss is thus provided by the relative length of the two sections of the E_o=constant line of Fig. 9.4b.

Consider now the particle motion in the velocity space as displayed in Fig. 9.5, which provides for the obvious relation

$$\frac{v_\parallel}{v_\perp} = \frac{\cos\theta}{\sin\theta}$$ (9.34)

or, after some trigonometric arranging

$$\sin^2 \theta = \frac{1}{\dfrac{v_{\parallel}^2}{v_{\perp}^2} + 1} . \qquad (9.35)$$

With the help of this expression, it becomes evident that Eq.(9.33) translates into the following statement: particles traveling in the weak magnetic field region, that is about the mid-plane, in directions such that

$$\sin \theta \ge \sqrt{\frac{B_{min}}{B_{max}}} , \qquad (9.36)$$

will be trapped, whereas the others moving within the so-called 'loss cone' with a polar angle

$$\theta_o = \arcsin \sqrt{\frac{B_{min}}{B_{max}}} \qquad (9.37)$$

will escape from the plasma region. In Eq.(9.36) only the positive sign of the square root is retained because θ can vary only between 0 and π. Note that Eq.(9.37) actually defines two values of θ_o which apparently relates to the two opposite loss cones thus including both the motions parallel and antiparallel to **B**. Denoting the one solution which is less than $\pi/2$ with θ_o', then the second solution is $\pi - \theta_o'$.

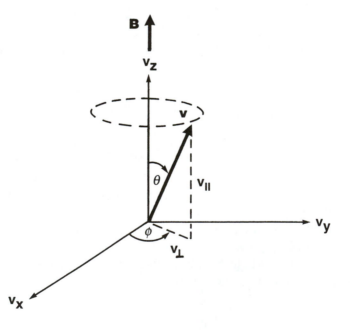

Fig. 9.5: Velocity space geometry with respect to the magnetic field direction. Note that θ = const. defines a cone on which the ratio $v_{\parallel} / v_{\perp}$ is invariant.

The fraction of an initially isotropic distribution of particles at the mid-plane, $f(\mathbf{v}) = f(v)/(4\pi v^2)$, which gets lost through both mirror ends, i.e. particles in both velocity cones, is then

$$f_{loss} = \frac{\displaystyle\int_{double\ cone} f(\mathbf{v})d^3v}{\displaystyle\int_0^\infty f(\mathbf{v})d^3v}$$

$$= \frac{\displaystyle\int_0^{2\pi} d\phi \left[\int_0^{\theta_o'} \sin\theta d\theta + \int_{\pi-\theta_o'}^{\pi} \sin\theta d\theta\right]\int_0^\infty \frac{f(v)}{4\pi v^2}v^2 dv}{\displaystyle\int_0^{2\pi} d\phi \int_0^{\pi} \sin\theta d\theta \int_0^\infty \frac{f(v)}{4\pi v^2}v^2 dv} \tag{9.38}$$

$$= 1 - \cos\theta_o' \ .$$

Evidently, then $\cos\theta_o'$ represents the fraction of the particle distribution which is trapped, that is

$$f_{trap} = \cos\theta_o' = \sqrt{1 - \frac{B_{min}}{B_{max}}} \tag{9.39}$$

where the ratio B_{max}/B_{min} is called the mirror ratio determining the effectiveness of confinement. As a consequence of the occurring end losses, a mirror plasma is never isotropic. Taking collisions into consideration, it is possible for particles which are in the confined region before collision, to be scattered into the loss cone.

So far we have discussed the confinement of charged particles by a magnetic mirror field which, however, could not explicitly explain why particles are reflected in the increased field of the mirrors. For that, we consider a particle moving outside the loss cone in velocity space. Its parallel motion can be stopped by the field strength of the mirror coils, that is the conserved total kinetic energy

$$E_o = E_\perp + E_\parallel$$
$$= \tfrac{1}{2}m(v_\perp^2)_{max} + 0 \tag{9.40}$$
$$= constant$$

is made up only by the perpendicular motion. If now, by some virtual means, the particle were displaced by an infinitesimal distance into the region of higher $\mathbf{B}$, then it would gyrate faster and thus increase its perpendicular velocity to $v_\perp > (v_\perp)_{max}$. This, however, would violate the energy conservation, Eq.(9.39), if no additional energy was transferred; it follows therefore, that the only directions the particle is allowed to move into–if it is displaced from the plane where it possesses $(v_\perp)_{max}$–are those associated with $v_\perp < (v_\perp)_{max}$ and thus making room again for $E_\parallel$. Since $v_\perp < (v_\perp)_{max}$ requires a reduced $\mathbf{B}$-field, the particle can regain

$E_\parallel$ only by moving opposite to ∇B and is thus reflected and returned to the weaker field region. Obviously, the force $\mathbf{F}_\parallel$ derived in Sec. 5.7 as $\mathbf{F}_\parallel = (\frac{1}{2}mv_\perp^2/B)\nabla_\parallel B$, Eq.(5.72), causes this mirror reflection.

9.4   Instabilities in Mirror Fields

In deriving the relative requirements of the magnetic field strength and the plasma pressure for effective confinement–Sec. 9.1–we considered, in the fluid model, the case of magnetohydrodynamic equilibrium. However, we did not examine whether this equilibrium state is stable or not. In an equilibrium state all forces are balanced allowing thus for a steady-state solution of the set of magnetohydrodynamic equations discussed at the end of Sec. 6.3. The equilibrium is labeled stable if small perturbations are inherently damped and it is unstable if small deviations from the equilibrium state are amplified, that is, if perturbations propagate and grow with time; this then is called an instability.

The unperturbed state requires perfect thermodynamic equilibrium in which the plasma particles have Maxwellian velocity distributions, while the plasma density and the magnetic field is uniform. Note that in the magnetic confinement configurations of interest to nuclear fusion reactors these requirements are not met. In the example of a mirror field, the isotropy of the Maxwell distribution is significantly disturbed by the particles lost through the mirror throats, which had obviously a dominant $v_\parallel$-component. Further, a mirror-device will feature ∇B and ∇N_i and thus be non-uniform. Though all forces can be balanced in a steady state, this state is not in perfect thermodynamic equilibrium and possesses so-called 'free' energy which can drive instabilities. Even periodic motions of the plasma fluid elements, e.g. plasma oscillations or, alternatively, waves, can thus be induced. An instability constitutes a motion which reduces the free energy and brings the plasma closer to perfect thermodynamic equilibrium. There exists a wide range and variety of possible plasma waves and instabilities, the discussion of which is far beyond the scope of this textbook. Hence we restrict ourselves to a few demonstrative examples of interest here.

The instability most relevant to mirror machines is the so-called flute-type instability. In the simple mirror geometry of Fig.9.3, the curvature of the magnetic field–except for the end-regions–is seen to be convex. Any outward perturbation of the confined plasma, i.e. a ripple on its boundary surface where all magnetically confined plasmas appear to have an energy density gradient, takes the plasma into regions of lower magnetic induction and lower kinetic pressure; hence, such a displacement to regions of reduced energy density will provide for free kinetic energy to let the perturbation grow. This can lead to flutes of plasma moving across magnetic field lines, Fig. 9.6, and result in particle loss from the containment region.

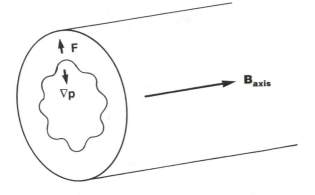

Fig. 9.6: Depiction of the flute instability.

Further insight into the onset and mechanism of the flute instability is provided if we recall the drifts and forces associated with a so-called "bad" convex **B**-field curvature, where the curvature drift, Eq.(5.61), will lead to charge separation occurring perpendicular to the magnetic field and the radius of curvature, Fig. 9.7. This polarization creates an azimuthal electric field causing an additional **E×B** drift, which transports both ions and electrons in the radially outward direction thus forming the flute-like bumps on the plasma column.

This type of instability can be avoided by generating a so-called 'minimum-B' field configuration in which the field lines are (almost) everywhere concave into the plasma. Here the charged particle then senses an increasing B-field in every direction and therefore finds itself in a magnetic well; the term minimum-B is thus commonly used for such a magnetic topology.

The simplest means of producing such a minimum-B field configuration is to locate four current carrying bars on the periphery of a magnetic device with their positions suggested in Fig. 9.8. These bars are called Ioffe bars with the current in adjacent bars flowing in opposite directions.

Another means of generating a minimum-B magnetic field configuration is by using a coil having the shape of the seam of a baseball. If the coil of the baseball configuration is suitably flattened and oriented in opposition with another similar coil, one obtains again a minimum-B configuration, called a "Yin-Yang" coil configuration, Fig. 9.9.

Further examination of these minimum-B configurations makes it clear that they all provide a central circular region for the plasma but that in two opposite directions the flattened fan-shaped magnetic fields are open and thereby still form a magnetic mirror. This feature can be extended by adding such devices to the ends of mirror solenoidal fields to form more effective mirrors because the magnetic well can in principle be deeper, and further, the particles contained therein will represent a thermal barrier to plasma escape.

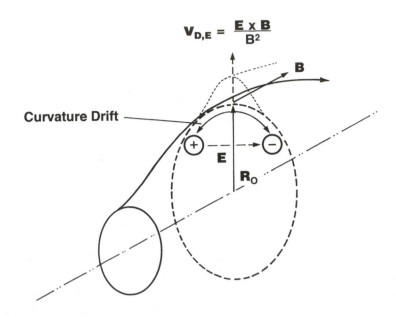

Fig. 9.7: Development of a flute instability: the curvature drift leads to azimuthal
polarization to create the **E**-field and so gives rise to the **E**✕**B** drift which displaces the
plasma particles radially outward.

Mirror devices are basically appropriate for steady state operation, in which
the particle injection rate balances the diffusion leakage rate. The diffusion
occurs dominantly through the open ends and constitutes also a diffusion in
velocity space, since–as previously mentioned–the velocity distribution in a
mirror plasma is no longer Maxwellian due to the preferred loss of particles with
large $v_{\parallel} / v_{\perp}$; one rather deals with a so-called loss-cone distribution excluding all
particles in the loss-cone as resulting from Eq.(9.38), Fig. 9.5. This deviation
from the Maxwellian distribution drives so-called velocity-space instabilities,
here specifically the 'loss-cone' instability, which can enhance the velocity-space
diffusion into the loss-cone. It has been observed that such instabilities are less
harmful to plasma confinement when the mirror device is short in dimension.

9.5   Classical Mirror Confinement

In the case that collisions are neglected, particles are trapped in a mirror when
they do not appear in the loss cone in velocity space. However, plasma ions and
electrons do suffer collisions, which can bring them randomly from the
confinement region of velocity space into the loss cone. Upon entering the loss

cone, the particles escape immediately within one transit time $\tau = L/v$ over the length L of the device. Due to their relatively small mass, electrons diffuse more rapidly both in velocity and coordinate space and hence are first to be scattered into the escape cone and then lost. This initially rapid loss rate causes the build-up of a positive electrostatic potential in the confined plasma which then consists of a surplus of ions having not yet scattered into the loss cone. Since this positive potential tends to retain the remaining electrons in the magnetic bottle, the overall plasma confinement time is governed by the ion escape time as characterized by the ion-ion collision time

$$\tau_{ii} = \frac{1}{N_i <\sigma_s v_r>} \tag{9.41}$$

and can be shown, for ions encountering simultaneous multiple collisions, to be given by

$$\tau_{ii} \propto \frac{A_i^{\frac{1}{2}} (kT_i)^{\frac{3}{2}}}{N_i q_i^4} \tag{9.42}$$

at the plasma temperatures of interest to nuclear fusion, with A_i representing the ion's atomic mass number.

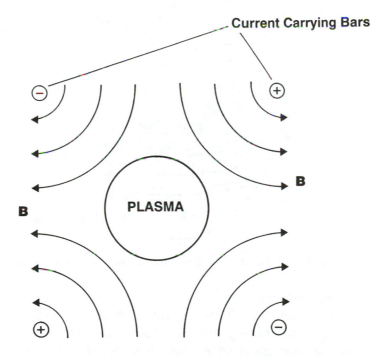

Fig. 9.8: Configuration of a minimum B-field.

The mirror confinement time τ_M must also be determined by the size of the loss cone, and consequently the approximation

$$\tau_M \approx \tau_{ii} \ln\left(\frac{B_{max}}{B_{min}}\right) \tag{9.43}$$

can be shown to hold for large values of the so-called mirror ratio B_{max} / B_{min}. By substituting, one obtains

$$\tau_M \approx C \frac{A_i^{\frac{1}{2}} T_i^{\frac{3}{2}} \ln\left(\frac{B_{max}}{B_{min}}\right)}{N_i Z_i^4} \tag{9.44}$$

where $q_i = Z_i e$ has been replaced and C is a constant taken for typical fusion temperatures here to be

$$C = 1.78 \times 10^{16} \text{ s} \tag{9.45}$$

when T_i is measured in keV, N_i in particles per m^3 and τ_M in seconds. Note that the mirror confinement time depends on the ion temperature and on the ratio B_{max}/B_{min}, but not on the actual magnitude of B or the plasma size. Evidently, a higher density will enhance the scattering into the loss cone and thus reduce τ_M.

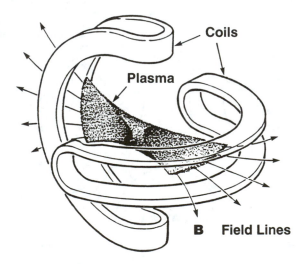

Fig. 9.9: Minimum B-field formed by a pair of Yin-Yang coils.

We add that the heuristic derivation of τ_M presented here refers to a classical treatment in the sense that collective effects such as instabilities are suppressed. Therefore, the above scaling applies when stabilization of those collective perturbations has been provided and then appears to be in good agreement with particle containment times recorded in magnetic mirror device experiments.

9.6 Magnetic Pinch

One of the simplest systems for magnetic containment is the pinch concept; here the plasma carries an electric current and is confined by the magnetic field induced by this current. As the current is increased, the larger magnetic field compresses the plasma and also raises its temperature by Joule-heating. Hence, confinement and heating is simultaneously provided. For that, extremely large currents (some 10^5 A) are needed thus rendering pinches to operate only in short pulses. The two principle configurations, denoted as z-pinch and θ-pinch, are sketched in Fig. 9.10.

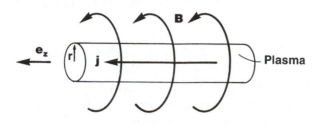

a) z - pinch

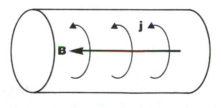

b) θ - pinch

Fig. 9.10: Representation of pinch geometries.

The confinement requirements are particularly obvious in the case of a z-pinch. In order that the plasma particles be retained, the electric currents must be high enough to generate a magnetic field of such strength that the magnetic energy density is able to balance the plasma pressure, that is

$$\frac{B^2}{2\mu_o} = (N_i + N_e)kT \tag{9.46}$$

where $T = T_i \approx T_e$ has been assumed. The magnetic field induced by $\mathbf{j}$ can be found from Maxwell's equation

$$\nabla \times \frac{\mathbf{B}}{\mu_o} = \mathbf{j} . \tag{9.47}$$

Using Stoke's integral theorem, we then relate

$$\oint \frac{\mathbf{B}}{\mu_o} \cdot d\mathbf{s} = \iint \left(\nabla \times \frac{\mathbf{B}}{\mu_o} \right) \cdot d\mathbf{A}$$

$$= \iint \mathbf{j} \cdot d\mathbf{A} \tag{9.48}$$

$$= \mathbf{I}$$

where ds is the path element along the circumference of the plasma column of cross section area **A** with d**A** denoting the oriented differential area element, and I defines the total electric current in the column. At the plasma surface, r = a, the path integral is

$$\int_0^a \frac{\mathbf{B}}{\mu_o} \cdot 2\pi dr \, \mathbf{e}_\theta = \frac{2\pi a}{\mu_o} B_\theta(a) \tag{9.49}$$

where $\mathbf{e}_\theta$ is the unit vector in the azimuthal direction.

Since **B** possesses only an azimuthal component, B_θ, the absolute value of the magnetic induction at the surface, is here given by

$$B(a) = \frac{\mu_o I}{2\pi a} . \tag{9.50}$$

Upon insertion of Eq. (9.50) in Eq.(9.46) and considering the particle numbers as referring to a plasma column of unit length, i.e. in the volume $a^2\pi \times 1$ m,

$$N^* = N_i^* + N_e^* = a^2\pi(N_i + N_e) , \tag{9.51}$$

we arrive at the requirement

$$\frac{\mu_o I^2}{8\pi} = N^* kT , \tag{9.52}$$

which is commonly referred to as the Bennett pinch condition.

Magnetic pinches are also troubled by instabilities. Two types of instabilities discussed here can occur in a cylindrical plasma carrying a large current along its axis.

One is the so-called 'sausage' instability, Fig. 9.11, which arises due to axial perturbations in the plasma column diameter and can be elucidated as follows. Consider the initial equilibrium state to be disturbed by some expansion of the plasma column to a larger radius. Since B(a) ~ 1/a, Eq. (9.50), the magnetic pressure at $a_1 > a_o$ is weaker than that at the equilibrium surface (a_o) and subsequently reinforces the expansion process. As these perturbations grow they make the plasma column look like a string of linked sausages and finally disrupt the plasma column. A restabilization of this pinch instability is possible by applying a sufficiently strong magnetic field along the axis.

Another characteristic instability in a cylindrical pinch is associated with perturbations of the linear axial geometry, i.e., when the plasma column exhibits an axial curvature as illustrated in Fig. 9.12. It is seen therein that the magnetic field lines are closer together at the inside of a helical bend and hence the magnetic induction is larger there than at the outside of the bend. This difference between the magnetic pressure at the inside and outside of the bend causes any small disturbance of the straightness of the equilibrium plasma cylinder to grow until the plasma column scrapes on the surrounding wall. This unstable behaviour against bending is called a helical kink instability and can as well be controlled by an additional strong axial magnetic field.

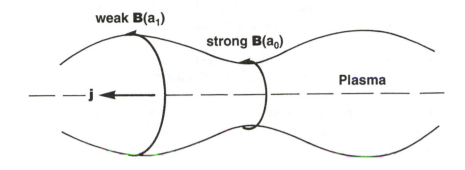

Fig. 9.11: Illustration of the sausage or pinch instability.

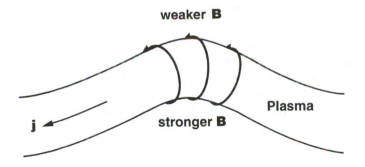

Fig. 9.12: Helical kink instability resulting from perturbations of the linear axial geometry.

Though a fusion device magnetically pinching the plasma is also seen to suffer instabilities which are mainly caused by the extreme electrical currents supplied, a θ-pinch can turn out to be remarkably stable. However, as for all open systems, linear pinches suffer from end leakage. To diminish end losses, a pinch usually features a magnetic mirror configuration, whereby also a minimum B-field structure is desirable. On the other hand, closed pinches can be realized when the plasma column is bent into a torus. Such a topology for magnetic confinement fusion will be discussed in the following chapter.

Problems

9.1   What assumptions are needed to make the previously stated magnetic confinement requirement, Eq. (4.12), consistent with Eq. (9.5)? Consider that the kinetic plasma pressure is largest at the centre and negligible at the plasma surface.

9.2   A magnetic mirror field, confining a fusion plasma with fuel particle densities $N_d = N_t = 4 \times 10^{19}$ m^{-3} at $T_i = 30$ keV, varies from $B_{max} = 6$ T at its throats to $B_{min} = 1$ T at the mid-plane. Find:
   (a) the number of ions escaping through the loss cones.
   (b) an approximate value of the mirror confinement time.

9.3   Depict the trapping and loss domains of deuterium and tritium ions along their E_o = constant lines, akin to those shown in Fig. 9.4, for the mirror configuration of problem 9.2.

9.4   Attempt to sketch a loss cone distribution function accounting for the absence of particles with large $v_{\parallel}/v_{\perp}$.

9.5   Discuss the development of the flute instability.

9.6   For each of the following open magnetic confinement fusion reactor concepts, draw a sketch of the concept, label the major components, and describe their purpose. Describe how fusion fuel ions are confined, and list what the energy and particle losses are in the system, and where they occur. Describe all the different electrical currents, magnetic fields, and their purpose. You may have to perform a literature search for some of the concepts.
   (a) Magnetic Mirror
   (b) Ying-Yang Coils
   (c) Z-pinch
   (d) Theta-Pinch
   (e) Reversed Field Mirror
   (f) Combined: Magnetic Mirror with Ying-Yang Coils

10.   Closed Magnetic Systems

Closed magnetic field configurations are those in which the field lines do not enter or leave the plasma confinement region, and thus offer the advantage of having no ends from which the plasma might escape. The simplest such configuration is a torus. Fusion devices containing toroidal plasmas have emerged as the dominant experimental facilities. We discuss here several of their distinguishing features as they relate to a conceivable energy generating device, emphasizing in particular, the tokamak concept.

10.1   Toroidal Fields

Bending a solenoidal coil around until its ends meet deforms an initially straight axial magnetic field into an axisymmetric toroidal field, as shown in Fig. 10.1. The toroidal field **B** is produced by simply passing a current through the coil wound around the torus. The torus is topologically defined by its major radius R_o and the minor radius a. The central line along the toroidal ring is called the minor axis, while the major axis is the one pointing out from the centre of the plane display of Fig. 10.1, i.e. going perpendicularly through the torus ring, Fig. 10.2.

Recalling our discussion of drift motion in Ch. 5, an examination of the toroidal configuration in Fig. 10.1 suggests that some complications do arise. The coils around the torus are more closely spaced on the inboard side than on the outboard side, creating therefore a radial variation in the magnetic field of the form

$$B(R) \propto \frac{1}{R} ,$$
(10.1)

as shown in Fig.10.2, with R denoting the distance from the major axis. Recall that a gradient in the magnetic field creates a grad-B force which drives the positive ions in one transverse direction and the negative electrons in the other. Further, the curvature drift drives them apart in the same way. The net result is a local charge separation, thereby generating a vertical electric field causing simultaneous ion and electron drift in an outward direction perpendicular to the major torus axis, Fig. 10.2.

To demonstrate these drifts we consider a simple magnetic field generated only from the current in the toroidal coils; then the field possesses only a component in the toroidal direction, $\mathbf{e}_\phi$, that is, we have

$$\mathbf{B} = \mathbf{B}_t$$

$$= (B_r, B_\phi, B_\theta) \tag{10.2}$$

$$= B_\phi \mathbf{e}_\phi \ .$$

The magnitude of the toroidal field can be calculated by employing Eq.(9.50) with I now representing the total current threading the hole of the torus. This then gives

$$B_\phi(R) = \frac{\mu_o I}{2\pi R} = \left(\frac{R_o}{R} \right) B_\phi(R_o) \tag{10.3}$$

and proves the proportionality introduced in Eq. (10.1). Considering a charged particle moving along $\mathbf{B}_t$ with speed $v_\parallel$, we note that it will be subjected to the curvature drift and, if it possesses also a velocity component perpendicular to $\mathbf{B}$, to the ∇B drift as well and hence, according to Eqs.(5.51) and (5.61), its guiding centre drift velocity ($\perp \mathbf{B}$) will be

$$\mathbf{v}_D = \frac{m}{q B^2} \left[v_\parallel^2 \frac{(\mathbf{R} \times \mathbf{B})}{R^2} + \left(\frac{v_\perp^2}{2} \right) \left(\frac{\mathbf{B}}{B} \times \nabla B \right) \right] . \tag{10.4}$$

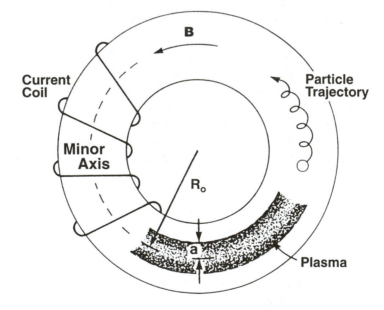

Fig. 10.1: A simple toroidal magnetic field configuration. Here, R_o is called the major radius and a is the minor radius.

Introducing Eqs.(10.2) and (10.3) into Eq.(10.4) we find the drift motion in the simple toroidal field here to be described by

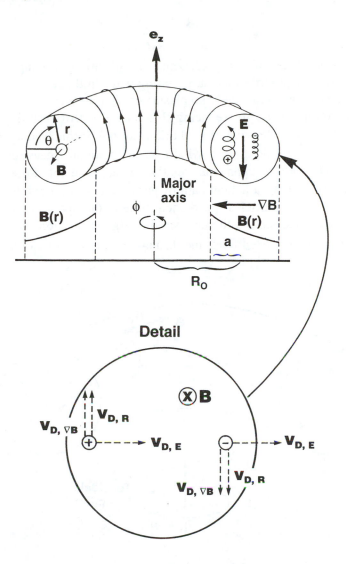

Fig. 10.2: Depiction of the radial dependence of the magnetic field and ion-electron separation in the torus.

$$\mathbf{v}_D = \frac{m}{q}\frac{1}{R_o\,B_\phi(R_o)}\left[\mathbf{v}_\parallel^2 + \frac{\mathbf{v}_\perp^2}{2}\right]\mathbf{e}_z \; . \qquad (10.5)$$

This equation is called the toroidal drift, the result of which is a polarization of the plasma with an electric field vector pointing in the negative z-direction. The resulting **E**-field unfortunately causes both the ions and electrons to move radially outwards due to the **E×B** drift with a velocity of

$$\mathbf{V}_{D,E} = \frac{\mathbf{E} \times \mathbf{B}}{B^2} = \frac{E}{B_\phi(R_o)} \cdot \frac{R}{R_o} \tag{10.6}$$

as shown in the detail of Fig. 10.2. Thus, the plasma in a simple toroidal field drifts outward until it strikes the surrounding wall. In this case, there is no radial equilibrium and hence practically no plasma confinement.

A common approach for reducing this outward drift resulting from charge polarization is the following. In addition to the toroidal magnetic field B_ϕ produced by the toroidal coil windings, a poloidal magnetic field B_θ is introduced. Depending upon the relative magnitude of these two **B**-field components, B_θ and B_ϕ, a helical path of particle migration results, as suggested in Fig.10.3. Then the particles spend equal time in the upper and lower halves of the toroid thus canceling out the undesired charge separation when the particles have moved along a twisted field line through a completed poloidal rotation, $\theta = 2\pi$.

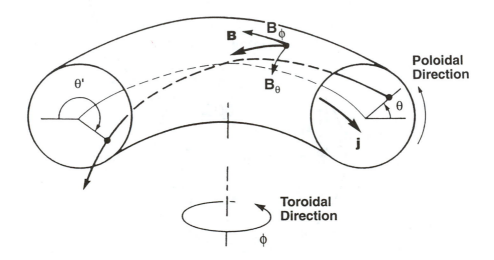

Fig. 10.3: Depiction of toroidal magnetic field, poloidal magnetic field, and rotational transform.

The poloidal angle made by a **B**-field line after it has traversed through a 2π-revolution in the toroidal direction (ϕ) is called the rotational transform, and represented by ι (iota). If ι is very large, the revolving field lines become too tight making the plasma unstable against kink-type perturbations, Sec. 9.6. Therefore it is useful to define a safety factor by

$$q = \frac{2\pi}{\iota} \tag{10.7}$$

which measures the field line pitch and it is seen from stability considerations to require q≥1 in the plasma and q≥2.5 at the edge of the plasma in order to suppress kink-type instabilities (Kruskal-Shafranov criterion). Thus, the rotational transform is constrained by

$$\iota < 2\pi \tag{10.8}$$

which means that, following a magnetic field line, its pitch has to be such that the number of revolutions around the major torus axis exceeds the number of revolutions around the minor axis.

While the rotational transform does provide for some spatial mixing of the charged particles and so greatly reduces the electric field and, consequently, the transverse outward drift, another interesting feature occurs. The effect of the twisted magnetic field lines–each of which completely traces out a magnetic flux surface by its revolutions around the toroidal and poloidal axes–is to create a system of nested toroidal flux surfaces which guide ion motion. Note, that in order to avoid the undesired charge separation, the rotational transform must not be an integer multiple of 2π. In that case, the field lines would recombine after some trips around the torus and hence not cover the entire magnetic flux surface on which they lie. Further, note that q may vary from one magnetic flux surface, as defined in Eq.(9.24), to the next, and this leads to a shear in the magnetic field since the **B**-lines point in a different direction as one proceeds radially onto different flux surfaces. We illustrate this in Fig.10.4. In typical tokamak plasmas q ranges from 1 near the centre of the plasma to values of 3-4 at the plasma edge. It is this shear which is effective against the growth of kink and drift perturbations. A perturbation aligned with $\mathbf{B}(r)$ will, at a point with increased minor radial distance r + dr, encounter field lines at a different angle which again will vary as the perturbation grows to another distance r + dr′. Any helically resonant instabilities are thus radially localized.

The poloidal magnetic field can be established in two ways. One is by arranging outside the plasma a set of poloidal coils which carry a current in the toroidal direction. Devices based on this principle are called stellarators. Others, known as tokamaks, employ a toroidal current flowing in the plasma; the latter happen to be relatively less complex devices and are therefore of greater current interest.

10.2 Tokamak Features

The most widely known toroidal fusion device is the tokamak; this word is of Russian origin and is a contraction of <u>to</u>roidal, <u>ka</u>mera (chamber), <u>ma</u>gnet and <u>k</u>atuschka (coil). The essential features of a tokamak are suggested in Fig.10.5, and the functions of the various components can be described as follows. The plasma torus is viewed as a single-winding secondary of a transformer. A current flow in the primary transformer winding will therefore induce a current in the plasma torus by transformer action. The resultant toroidal plasma current I provides for Ohmic heating and also generates the poloidal magnetic field B_θ. Further, the toroidal magnetic field coils generate the toroidal field B_ϕ which is perpendicular to B_θ. These two magnetic fields combine by vector addition,

$$B = B_\phi + B_\theta \qquad (10.9)$$

to provide the basis for the rotational transform in Fig. 10.3. Further, the stabilizing coils, Fig.10.5, are to restrain the plasma from expanding as is the natural tendency of a current ring. Then, the total field vector **B** has to incorporate also this stationary field component. The expansion tendency arises from the difference between the poloidal magnetic pressure, $B_\theta^2 / 2\mu_o$ (Sec. 9.1), on the outside and the inside of the torus. Since B_θ is larger on the inside, the associated higher magnetic pressure will steer the plasma ring towards an increased major radius. It is therefore seen that toroidal and poloidal magnetic fields alone cannot confine the plasma in a tokamak. Of additional need is a vertical magnetic field B_v, produced by the stabilizing coils shown in Fig. 10.5, in order to produce an inward directed $j \times B_v$ force which prohibits the outward expansion. If the toroidal plasma current is sufficiently strong, only a small B_v is required to stabilize the plasma. Hence, a tokamak may be characterized as a toroidal device featuring a large plasma current and a strong toroidal magnetic field such that

$$B_\phi > B_\theta > B_v . \qquad (10.10)$$

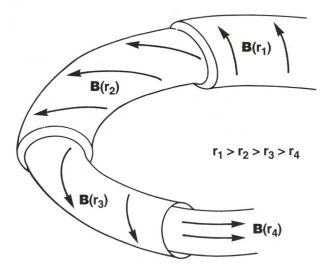

Fig. 10.4: Nested magnetic flux surfaces with different shear.

The toroidal plasma current needed to supply the poloidal magnetic induction is generated by a toroidal electric field $\mathbf{E} = (0, E_\phi, 0)$ which is achieved here by the transformer's temporally changing magnetic flux Ψ_{trans} that penetrates the hole in the torus.

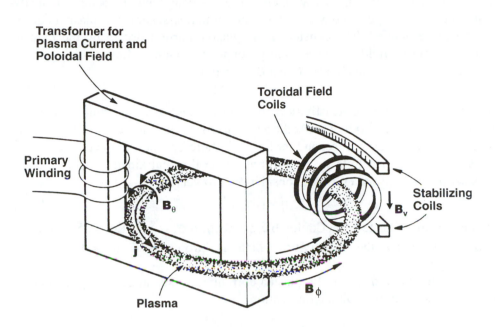

Fig. 10.5: Components of a tokamak.

The connection between the various physical field quantities involved is readily evident from the following. Since one of Maxwell's Equations tells us

$$\nabla \times \mathbf{E} = -\frac{d\mathbf{B}}{dt} , \qquad (10.11)$$

we conclude from Stoke's Law,

$$\oint \mathbf{E} \cdot d\mathbf{s} = \iint_A (\nabla \times \mathbf{E}) \cdot d\mathbf{A} = -\iint_A \frac{d\mathbf{B}}{dt} \cdot d\mathbf{A}$$

$$= -\frac{d}{dt} \int_{A=const} \mathbf{B} \cdot d\mathbf{A} = -\frac{d\Psi_{trans}}{dt} , \qquad (10.12)$$

that the work done per unit charge by the induced electric field over a closed path along the secondary transformer winding–which essentially is the plasma ring itself–is equal to the rate of change of magnetic flux, Ψ_{trans}, through the closed winding loop, i.e. through the hole in the torus. Hence, $d\mathbf{s} = 2\pi \, d\mathbf{R} \, \mathbf{e}_\phi$ here and $A = \pi R^2 = $ constant, which has been assumed in Eq.(10.12). The induced toroidal electric field is then found from Eq.(10.12) to be of the magnitude

$$E_\phi = -\frac{1}{2\pi R}\frac{d\Psi_{trans}}{dt}. \qquad (10.13)$$

The flux change is achieved either by an air-core or an iron-core transformer as shown in Fig. 10.5. It follows then that a tokamak cannot operate in steady-state but only as a pulsed device unless non-inductive schemes of plasma current drive can be applied. The density of the plasma current driven by the induced toroidal electric field is subsequently found via Ohm's Law which can be generalized for a conducting fluid in a **B**-field to be

$$\eta \mathbf{j} = \mathbf{E} + \mathbf{V} \times \mathbf{B} \qquad (10.14)$$

where η is the specific resistivity of the plasma and **V** is the fluid velocity as defined and used in Eq.(6.21). Additional current terms can occur in Eq.(10.14), which, however, are of second order and therefore not itemized here. The resistivity of a thermal plasma is governed by collisional kinetic effects and can be shown from a plasma physics consideration to vary as

$$\eta \propto (kT_e)^{-3/2}. \qquad (10.15)$$

As the plasma is heated, the Coulomb cross section $\sigma_s \propto v_r^{-4}$ (Eq. 3.15) decreases and consequently the resistivity, which is proportional to $<\sigma_s v_r>$, drops accordingly.

Note that the induction of a plasma current suggests an easy way to heat the plasma; that is through Ohmic dissipation

$$P_{OH} = \int_{Volume} \eta\, j^2 d^3 r, \qquad (10.16)$$

where the symbol OH refers to the common label of this effect. It is evident from the proportionality given in Eq.(10.15) that this heating method becomes less efficient at higher plasma temperatures (>1 keV) and will not suffice up to the required thermonuclear temperatures (~10 keV). For that, other methods must be employed of which some we discussed in Sec. 8.2.

Another effect associated with electric field induction in a plasma is noted here. The frequency of collisions between electrons and ions (compare with $1/\tau_{ii}$ of Eq.(9.41)) is

$$(\tau_{ei})^{-1} \propto \sigma_s v_r \propto v_r^{-3} \qquad (10.17)$$

and shows that high-energy electrons undergo relatively few collisions and therefore predominantly carry the induced current. Consider now an electron from the high energy tail of its velocity distribution which moves in the direction opposite to **E**. Due to the low collision frequency, it will gain further energy making thus a collision with an ion even less likely. This in turn allows it to be further accelerated by **E**. This phenomenon is called 'electron runaway'. If E, and hence the velocity gain of the electrons, is sufficiently large, the Coulomb cross section drops so quickly that these runaway electrons never encounter a collision and thus form a beam of accelerated electrons disengaged from the main part of the distribution.

10.3 Particle Trapping

In an axisymmetric case, the flux surfaces (Fig. 10.4) embracing the magnetic field lines are, to a good approximation, annular toroidal surfaces with r = constant. The charged particles follow the helical field lines resulting from the combination of $\mathbf{B}_\phi + \mathbf{B}_\theta$ and hence move on the flux surfaces, except for excursions of the order of the gyroradius.

While most of the particles are free to spiral around the helical field lines as they encircle both the major and minor axis of the torus, there is a class of particles which appear to be trapped in a magnetic well formed by the field variation between the inboard and outboard side of the torus. Both the toroidal and poloidal fields are stronger on the inside than on the outside of the torus, which results in an overall field variation as illustrated in Fig. 10.6 and at length amounts to a sequence of magnetic mirrors. Some of the plasma particles exhibiting a lower $v_\parallel$ in comparison with the particles moving completely about the torus (known as passing particles), are trapped by these mirrors according to the effects discussed in Ch. 9. If the particle trajectory from a number of toroidal journeys is projected onto a transverse plane of the torus (ϕ = const), a kind of banana-shaped orbit results for trapped particles, i.e. for particles encountering the mirror reflection. This is displayed in Fig. 10.7, where the trajectory of an untrapped particle is also illustrated. Its guiding centre motion is seen on a curve not quite coinciding with the corresponding magnetic flux surface, which is due to first order drift motion across $\mathbf{B}$.

Recalling from Chapter 9 that the fraction of particles trapped in a mirror field is

$$f_{trap} = \sqrt{1 - \frac{B_{min}}{B_{max}}} \qquad (10.18)$$

and assuming $B_\theta \ll B_\phi$ such that

$$B \approx B_\phi \propto \frac{1}{R_o + r\cos\theta}, \qquad (10.19)$$

we find here the ratio

$$\frac{B_{min}}{B_{max}} = \frac{B_{outboard}}{B_{inboard}} \approx \frac{B(r=a)}{B(r=-a)}$$

$$= \frac{\dfrac{1}{R_o + a}}{\dfrac{1}{R_o - a}} = \frac{1-\varepsilon}{1+\varepsilon} \qquad (10.20)$$

where $\varepsilon = a/R_o$ is the inverse aspect ratio of the tokamak. Thus, the fraction of trapped particles in a tokamak is given by

$$f_{trap} \approx \sqrt{1 - \frac{1 - \varepsilon}{1 + \varepsilon}} = \sqrt{\frac{2\varepsilon}{1 + \varepsilon}}. \tag{10.21}$$

For a tokamak reactor with $a/R_o \approx 1/3$, approximately 70% of the plasma particles appear to be trapped, and thus their enhanced radial diffusion across the confining magnetic field is viewed to be a significant process in tokamak particle leakage.

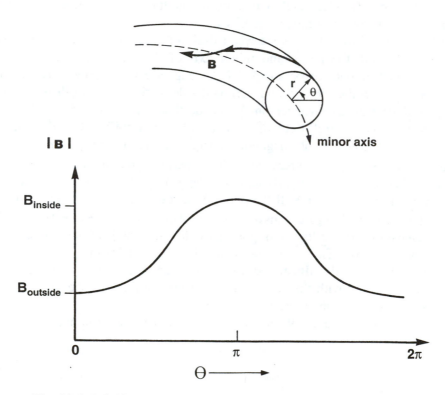

Fig. 10.6: Poloidal variation of the overall magnetic field in a tokamak.

The trapped particles bounce back and forth between poloidal angles of $\pm\theta_b$, thus avoiding the inner torus region. In combination with the toroidal drift this bounce motion results in the poloidal cross sectional orbit shown in Fig. 10.7, from where the name 'banana orbit' becomes self evident.

Apparently, the poloidal rotation of the magnetic field lines tends to suppress the vertical drifts due to field curvature and ∇B for trapped particles as well; they would spend equal times in the upper and lower halves of the torus if there were no toroidal electric field which, however, is necessary for driving the plasma current. As a consequence of this E-field, the banana orbits are no longer symmetric. It is seen that an ion spends a longer time period in the lower half of the torus drifting radially inward than in the upper half where it drifts radially outward. In total, the trapped particles should be subjected to a net inward drift towards smaller minor radii, which thus pinches the plasma column. This phenomenon was predicted by A. Ware and is since called the Ware-pinch effect.

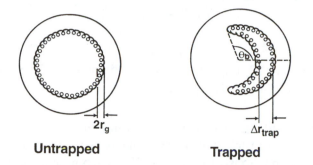

Untrapped   **Trapped**

Fig. 10.7: Schematic projection of particle orbits in a tokamak onto a poloidal plane. Here Δr_{trap} indicates the banana width of the trapped particle orbit.

Another effect associated with banana trapping is the enhanced particle leakage. As previously discussed, collisions can cause particles to jump across the confining magnetic flux surfaces and thereby determine the diffusion losses. While the maximum distance which an untrapped particle can be displaced as a result of a single collision is of the order of the gyration radius r_g, a trapped particle can be subjected to a maximum excursion as great as the banana width Δr_{trap} indicated in Fig. 10.7. Since typically $\Delta r_{trap} > r_g$ and the co-efficient for diffusion perpendicular to the magnetic field, D, increases with $r_g^2 = \overline{v}^2 / \omega_g^2$ (Eq.(6.16b)) and Δr^2_{trap}, respectively, the trapped particles may escape more rapidly from the tokamak plasma than the untrapped ones. Evidently, if the fraction of trapped particles is large, this leakage enhancement constitutes a substantial problem in tokamak confinement.

10.4   Tokamak Equilibrium

An equilibrium state of the plasma confined by magnetic fields must exist in order to maintain the hot plasma away from the vessel walls. While the general solution of equilibrium states of a magnetically confined plasma constitutes a complicated problem, some useful physical relations follow readily from viewing the plasma as a single fluid (Sec. 6.3) subjected to the magnetohydrodynamic (MHD) description. The equilibrium exploration suggests that one consider the steady state MHD equations from which the plasma current density **j** and the magnetic field **B** can be self-consistently derived, i.e., in agreement with Maxwell's Equations. Consequently, an equilibrium plasma must satisfy the previously introduced force balance (Eq.(9.2))

$$\nabla p = \mathbf{j} \times \mathbf{B} \tag{10.22}$$

and the magnetostatic Maxwell equations

$$\nabla \times \mathbf{B} = \mu_o \mathbf{j} \tag{10.23}$$

and

$$\nabla \cdot \mathbf{B} = 0 . \tag{10.24}$$

As in Sec. 9.1, Eq. (10.22) implies that **j** and **B** lie on isobaric surfaces. Note that taking the divergence of Eq. (10.23) yields

$$\nabla \cdot \mathbf{j} = 0 \tag{10.25}$$

which corresponds with the quasi-neutrality assumption $N_i \approx N_e$.

For an axisymmetric tokamak plasma the isobaric surfaces will be nested toroidal surfaces, as previously mentioned. We assign now to each such surface a function Ψ = constant such that $2\pi\Psi$ represents the poloidal magnetic flux through a toroidal plane extending from the minor axis (r = 0) to the considered isobaric surface at r represented by the shadowed area in Fig. 10.8, i.e.

$$2\pi\Psi := \int \mathbf{B}_\theta \cdot d\mathbf{r}_\theta . \tag{10.26}$$

Since isobaric surfaces need not necessarily have circular cross sections, here r_θ denotes the radial distance of an isobaric surface at a specified poloidal position θ. Then we may approximate the flux passing through the planar ribbon between the two isobaric surfaces labeled Ψ and $\Psi + d\Psi$ and encircling the major axis with a radius R by

$$2\pi\left(\Psi + d\Psi\right) - 2\pi\Psi = \int_0^{r_\theta(\Psi+d\Psi)} \mathbf{B}_\theta \cdot d\mathbf{r}_\theta - \int_0^{r_\theta(\Psi)} \mathbf{B}_\theta \cdot d\mathbf{r}_\theta \approx B_\theta 2\pi R dr \tag{10.27}$$

with dr representing the radial distance between the considered surfaces. From Eq.(10.27), the following relation between the poloidal magnetic field strength and the flux surface function Ψ is now apparent:

$$RB_\theta = \frac{d\Psi}{dr} \tag{10.28}$$

or, respectively, in vectorial form according to the orientation shown in Fig.10.8,

$$\mathbf{B}_\theta = -\frac{1}{R}\mathbf{e}_\phi \times \nabla\Psi . \tag{10.29}$$

Similarly, due to the symmetry of **j** and **B** in Eq. (10.22), there exists a corresponding poloidal current function $F(\Psi)$ such that the poloidal component of the current density can be expressed as

$$\mathbf{j}_\theta = -\frac{1}{R}\mathbf{e}_\phi \times \nabla F \tag{10.30}$$

which–via Eq. (10.23)–determines the toroidal field as

$$\mathbf{B}_\phi = -\mu_o \frac{F}{R} \mathbf{e}_\phi \ . \tag{10.31}$$

Since F = constant for Ψ = constant, it is therefore also seen that $B_\phi \times R$ is constant on a magnetic surface.

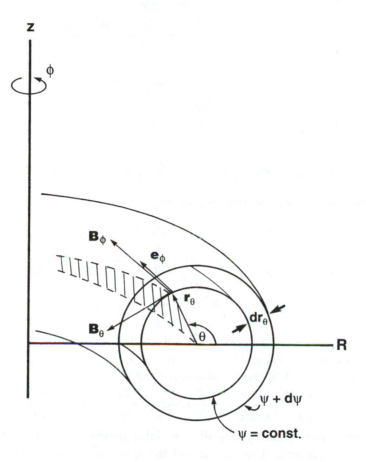

Fig. 10.8: Toroidal and poloidal field components in a cylindrical coordinate system, also indicating the nested flux surfaces Ψ = constant.

Next, we decompose the poloidal magnetic field into its components along the R and z axes of Fig. 10.8 by multiplying Eq.(10.29) with the respective unit vectors $\mathbf{e}_R$ and $\mathbf{e}_z$. Noting that

$$\mathbf{e}_R \times \mathbf{e}_\phi = \mathbf{e}_z \tag{10.32}$$

we find

$$B_R = -\frac{1}{R} \frac{\partial \Psi}{\partial z} \tag{10.33a}$$

and

$$B_z = \frac{1}{R}\frac{\partial\Psi}{\partial R} \; . \qquad\qquad (10.33b)$$

Using these expressions the toroidal component of Eq. (10.23) is then

$$\mu_o \mathbf{j}_\phi = \left(\nabla\times\mathbf{B}(B_R,B_\theta,B_z)\right)_\phi = -\frac{\partial}{\partial R}B_z + \frac{\partial}{\partial z}B_R$$

$$= -\frac{\partial}{\partial R}\frac{1}{R}\frac{\partial\Psi}{\partial R} - \frac{1}{R}\frac{\partial^2\Psi}{\partial z^2} \qquad (10.34)$$

which defines the operator

$$\Delta^*\Psi = R\frac{\partial}{\partial R}\frac{1}{R}\frac{\partial\Psi}{\partial R} + \frac{\partial^2\Psi}{\partial z^2} \qquad (10.35)$$

used in the following. Upon introducing the flux and the current function in the force balance, Eq. (10.22),

$$\nabla p = \left(\mathbf{j}_\phi + \mathbf{j}_\theta\right)\times\left(\mathbf{B}_\phi + \mathbf{B}_\theta\right) = \left(\mathbf{j}_\phi - \frac{1}{R}\mathbf{e}_\phi\times\nabla F\right)\times\left(\mathbf{B}_\phi - \frac{1}{R}\mathbf{e}_\phi\times\nabla\Psi\right), \quad (10.36a)$$

and recognizing the pertinent orthogonalities, we obtain

$$\nabla p(\Psi) = \frac{j_\phi}{R}\nabla\Psi - \frac{B_\phi}{R}\nabla F(\Psi) . \qquad (10.36b)$$

Since $\nabla p(\Psi) = \dfrac{dp}{d\Psi}\nabla\Psi$ and $\nabla F(\Psi) = \dfrac{dF}{d\Psi}\nabla\Psi$, Eq.(10.36b) provides the relation

$$j_\phi = R\frac{dp}{d\Psi} + B_\phi\frac{dF}{d\Psi} = R\frac{dp}{d\Psi} + \frac{\mu_o}{R}F(\Psi)\frac{dF}{d\Psi} \qquad (10.37)$$

where, in the last term, B_ϕ has been substituted by Eq.(10.31). Introducing Eq.(10.37) into Eq.(10.34) and using the operator definition of Eq.(10.35) yields finally the famous Grad-Schlueter-Shafranov Equation

$$\Delta^*\Psi = \frac{\partial^2\Psi}{\partial R^2} - \frac{1}{R}\frac{\partial\Psi}{\partial R} + \frac{\partial^2\Psi}{\partial z^2} = -\mu_o R^2\frac{dp}{d\Psi} - \mu_o^2 F(\Psi)\frac{dF}{d\Psi} \qquad (10.38)$$

which is a nonlinear elliptic partial differential equation for the magnetic flux Ψ in terms of the pressure distribution and the poloidal current function F. The corresponding boundary conditions are provided by the transformer-induced poloidal magnetic field outside the plasma. In evaluating Eq. (10.38) it is seen and has been discussed earlier that the total magnetic field must also possess a uniform vertical part $\mathbf{B}_v$ in order to prevent the current-carrying plasma from radial expansion; only then can a toroidal plasma be maintained in equilibrium. The topological magnetic field structure effected by the imposition of $\mathbf{B}_v$ is demonstrated in Fig. 10.9 where, in the resulting field, the domain of closed magnetic surfaces is separated by a so-called separatrix from the open surface contours intercepting the wall of the reactor chamber.

In practice, the Grad-Schlueter-Shafranov Equation, Eq.(10.38), is solved numerically to find the geometrical location of the magnetic surfaces and the radial distribution of the axial current density in a way that is consistent with the experimentally measured pressure profiles and the externally applied field. That is, the functions p(Ψ) and F(Ψ) are given for the numerical evaluation. For plasma pressures which are not too high, solutions of Eq.(10.38) exist in the form of closed nested flux surfaces inside the separatrix indicated in Fig.10.9.

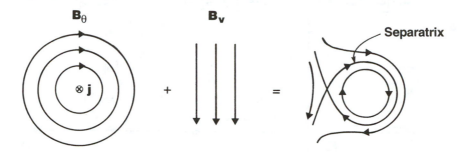

Fig. 10.9: Effect of superimposing a uniform vertical magnetic field onto a poloidal field B_θ induced by a plasma current **j** flowing into the plane of the page.

Various equilibria become apparent in a tokamak plasma and are characterized by the so-called poloidal beta

$$\beta_p = \frac{<p>}{B_\theta^2(a)/2\mu_o} \tag{10.39}$$

which is the ratio of the volume-averaged plasma pressure to the energy density of the poloidal magnetic field averaged over the magnetic flux surface at r = a. Figure 10.10 exhibits some typical features of toroidal equilibria as they are affected by the plasma pressure. A more detailed analysis shows that the toroidal field contributes to the pressure balance by the difference $\left\langle B_\phi^2 \right\rangle_{\Psi(a)} - \langle B_\phi^2 \rangle$ affected by the poloidal current components which may either increase or decrease the toroidal field in the plasma. A plasma is paramagnetic when the volume-averaged square of the toroidal field strength, $\langle B_\phi^2 \rangle$, exceeds the outer flux surface average, $\left\langle B_\phi^2 \right\rangle_{\Psi(a)}$, and it is diamagnetic for $\left\langle B_\phi^2 \right\rangle_{\Psi(a)} > \langle B_\phi^2 \rangle$. As a consequence, the diamagnetic plasma is associated with an average plasma pressure greater than $\left\langle B_\theta^2 \right\rangle_{\Psi(a)} / (2\mu_o)$, that is with $\beta_p > 1$, while $\beta_p < 1$ corresponds to a paramagnetic plasma. Whereas the total beta of a tokamak plasma is constrained to values of a few percent for MHD stability reasons, the poloidal beta appears in the range $0.1 \leq \beta_p \leq 2.5$ for the equilibrium regimes displayed in Fig. 10.10.

It is observed that the magnetic axis of the torus column is displaced outwards with increasing plasma pressure until this displacement D is about half of the minor plasma radius, which can be shown to manifest the limit of

Surfaces on which p = constant

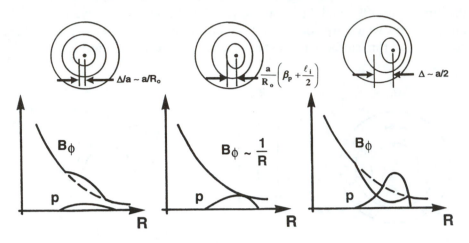

Surfaces on which j_ϕ = constant

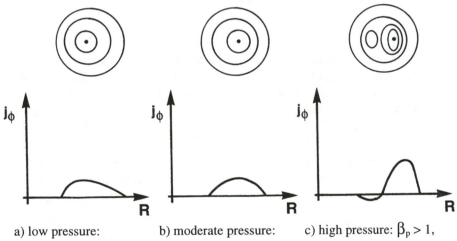

a) low pressure:

$\beta_p < 1$, paramagnetic
behaviour of plasma
increases toroidal field

b) moderate pressure:

$\beta_p \approx 1$, toroidal field is
almost equal to the
1/R - vacuum field

c) high pressure: $\beta_p > 1$,

diamagnetic behaviour of
plasma decreases toroidal field;
occurrence of reverse current

Fig. 10.10: Axisymmetric tokamak equilibrium characteristics varying with plasma
pressure. The symbol ℓ_i appearing in b) denotes the plasma internal inductance per unit

$$\text{length } \ell_i = \frac{\int\limits_0^{2\pi} d\theta \int\limits_0^a B_\theta^2 r dr}{\pi a^2 \langle B_\theta^2 \rangle_{\Psi(a)}} \; .$$

equilibrium. The poloidal magnetic field must be capable of supporting at least the fraction a/R_o of the kinetic pressure, that is

$$\frac{B_\theta^2(a)}{2\,\mu_o} > \frac{a}{R_o} <p> \tag{10.40a}$$

which obviously restricts β_p to values

$$\beta_p < R_o / a . \tag{10.40b}$$

Recalling the Kruskal-Shafranov stability limit in Sec.10.1 which suggests that the safety factor at the plasma edge should be $q(a) > 2.5$, and further extracting from the geometric sketch in Fig. 10.11 the obvious relation

$$\tan \delta = \frac{B_\theta(a)}{B_\phi} \approx \frac{a\Delta\theta}{R_o\Delta\phi} \tag{10.41}$$

with δ denoting the pitch angle, we find the requirement

$$\frac{B_\phi}{B_\theta(a)} \approx q(a)\frac{R_o}{a} > 2.5\frac{R_o}{a} \tag{10.42}$$

by using the previous definitions $\iota = \Delta\theta(\Delta\phi = 2\pi)$ and $q = 2\pi\iota$. Taking the square of Eq.(10.42) and assuming the approximate equivalence of $B_\theta^2(a)$ with $\left\langle B_\theta^2\right\rangle_{\Psi(a)}$ and of B_ϕ^2 with $\left\langle B_\phi^2\right\rangle_{\Psi(a)}$, we express the averages of both $B_\theta{}^2(a)$ and $B_\phi{}^2$ by means of their β-value to obtain

$$\frac{<p>}{B_\phi^2 / 2\,\mu_o} := \beta_t \approx \frac{a^2}{R_o^2}\frac{\beta_p}{q^2(a)} \leq \frac{1}{6.25}\frac{a^2}{R_o^2}\beta_p \tag{10.43}$$

which, upon introduction of the conclusion drawn from Eq. (10.40b) yields

$$\beta_t < \frac{a}{R_o}\frac{1}{q^2(a)} \leq \frac{1}{6.25}\frac{a}{R_o} . \tag{10.44}$$

Neglecting the inferior magnetic pressure of the vertical field B_v, the total plasma-beta

$$\beta = \frac{<p>}{\dfrac{B_\theta^2(a)}{2\,\mu_o} + \dfrac{B_\phi^2}{2\,\mu_o}} \tag{10.45}$$

is then seen to be constrained to values

$$\beta < \frac{a / R_o}{q^2(a) + \dfrac{a^2}{R_o^2}} \leq \frac{a / R_o}{6.25 + \dfrac{a^2}{R_o^2}} \tag{10.46}$$

which essentially is the toroidal beta, β_t, for large aspect ratios. Hence, for example, a tokamak featuring an aspect ratio $R_o/a \approx 3$ will be limited to $\beta < 5\%$ by principal MHD stability considerations, while the poloidal beta may amount to values in the order of R_o/a.

Another important consequence resulting from the poloidal-beta limitation becomes evident when the poloidal magnetic induction is expressed–via Ampere's Law–by the plasma current generating the B_θ-field. Doing so, we introduce Eq. (9.50) into Eq. (10.42) to obtain

$$\frac{B_\phi}{\dfrac{\mu_o I}{2\pi a}} > q(a)\frac{R_o}{a} \tag{10.47}$$

and subsequently the constraint

$$I < \frac{2\pi}{\mu_o}\frac{a^2}{R_o q(a)}B_\phi \ . \tag{10.48}$$

That is, the plasma current cannot be increased arbitrarily in order to achieve higher temperatures through Ohmic heating. Its magnitude appears to be bounded for stability reasons. We inject to note that the above derivation of plasma-beta constraints associated with simple MHD stability limits represents just a rough assessment nonetheless demonstrating the useful and important relations in an illustrative manner. Thorough analysis would provide more accurate factors in the several inequalities at the expense of extended expressions and more detailed concepts to be introduced.

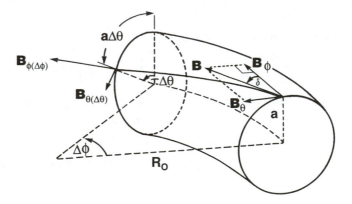

Fig. 10.11: Toroidal and poloidal magnetic induction vectors varying with toroidal and poloidal rotation.

10.5   Stability of Tokamaks

Considering plasma states which are not in perfect thermodynamic equilibrium (no exact Maxwellian distribution), even though they represent equilibrium states in the sense that the force balance is equal to 0 and a stationary solution exists, means their entropy is not at the maximum possible and hence free energy appears available which can excite perturbations to grow. Such an equilibrium state is unstable. The stability of a plasma confined by a toroidal and poloidal magnetic configuration is therefore seen to be determined by the free energies associated with eventual currents parallel to **B** (force-free currents) and with the plasma pressure. The ratio of these two free energies turns out to be β_p. Apparently, the gradients of plasma current magnitude and pressure, ∇j and ∇p, are the destabilizing forces in connection with the 'bad' magnetic field curvature discussed in Sec. 9.4 and necessarily inherent to a tokamak.

The occurring instabilities can be divided into three types:

(a) Ideal MHD Modes:

Amongst all types, the ideal MHD instabilities are the most virulent due to their fast growth and the possible extension over the entire plasma. Thus, on the short time scale of relevance (microseconds) the resistivity of the plasma is negligible.

A toroidally confined plasma sees 'bad' convex curvature of the helical magnetic field lines on the outboard side of the torus. Consequently, such a configuration is subject to the onset of flute-type interchange instabilities of which the driving mechanisms have already been demonstrated in Sec. 9.4. However, because the magnetic field lines in a tokamak are more concentrated on the inboard side where there is 'good' curvature (concave as seen from the plasma), the average curvature of **B**-field lines over a full poloidal rotation is 'good' for windings with a rotational transform $\iota \le 2\pi$, i.e., $q \ge 1$. Therefore, interchange perturbations do not grow in normal ($q \ge 1$) tokamaks. It is however observed that the perturbations can locally grow or 'balloon' in the outboard 'bad' curvature region. In that case a high local pressure gradient is responsible for driving the so-called ballooning instability. By establishing appropriate pressure profiles and appropriate magnetic field line windings, those modes can be suppressed almost everywhere in the plasma.

Another instability which represents the most dramatic one in ideal MHD is the kink instability already discussed in Sec. 9.6. It causes a contortion of the helical plasma column and consequently of the magnetic flux surfaces. Preferably it occurs in tokamak plasmas at low pressures and is driven by the radial gradient of the toroidal current. Fortunately these instabilities are bounded to small intervals of q(a) lying close below integer values. For the current profile distributions typical of tokamak fusion plasmas, such unstable modes arise mainly when q(a) is just a little less than 2. The associated kink distortion of the plasma column can be stabilized by an enhanced toroidal magnetic field strength such that the Kruskal-Shafranov condition

$$\frac{B_\theta}{B_\phi} < \frac{a}{R_o} \tag{10.49}$$

is fulfilled, which we had used previously and extended to a more stringent criterion at the plasma edge, Eq. (10.42).

As seen in the preceding section, adjusting the safety factor q to the appropriate value is associated with a limitation of the plasma beta. In order to avoid the major MHD unstable activities the overall β is limited by the maximum critical beta

$$\beta_{crit} [\%] = C_T \frac{I}{aB} \qquad (10.50)$$

with the so-called Troyon factor C_T (dimensionless) ranging from 2.8 to 5 depending on the nature of the instabilities, if the plasma current I is in MA, the minor radius a in m and the confining field in Tesla. Since the plasma cross section need not necessarily be of circular shape (actually, most tokamak plasmas of today's experiments feature an elliptic, bean or D-shaped cross section which permit optimization of energy confinement and plasma pressure profiles), parameters which account for the actual cross-sectional contour will also enter the Troyon factor. Note that the rough analysis in the preceding section, Eqs.(10.40) - (10.48), can also provide for the functional relation given in Eq. (10.50).

(b) Resistive MHD Modes

Additional types of macroscopic instabilities are attributable to the electrical resistivity of a tokamak plasma, which makes the instability grow more slowly. Characteristically, the growth times are of the order of 10^{-4} to 10^{-2} s which, however, is still short compared with the energy confinement time τ_E (seconds). Resistive MHD modes result from the diffusion or tearing of magnetic field lines relative to the plasma fluid and can thereby destroy the nested topology of the magnetic flux surfaces. For helically resonant **B**-perturbations, magnetic field diffusion may preponderate the ideal MHD effects in thin boundary layers around surfaces having a rational q. Then the magnetic field lines can reconnect in these layers thereby producing nonaxisymmetric helical islands as suggested in Fig. 10.12. The tearing mode instability in a tokamak is driven by the radial gradient of the equilibrium current density. Upon formation of a magnetic island filament it grows until it acquires all the accessible free energy of the current. Outside these so-called resonant surfaces (q is rational) the plasma undergoes a sequence of MHD equilibria.

For low-β plasmas it is possible that these resistive modes couple nonlinearly amongst each other (different rational q), which leads to a disruption of the plasma current. Increasing the plasma current, which–according to Eq. (10.48)–is associated with lowering the safety factor q, is seen to diminish the existence of current distributions which are stable against tearing modes.

Another effect brought about by the nonlinear evolution of resistive MHD modes and observed in tokamak plasmas is the 'sawtooth' behaviour of some plasma parameters such as the electron temperature and the current density at the centre of the plasma column where q can drop below 1. Sawtooth oscillations can be delayed or prevented by modifying the current profile near the q = 1 magnetic surface. These sawtooth oscillations are not deemed catastrophic since their activity is constrained within a small region internal to the q = 1 surface, where it comes to a redistribution of the plasma energy.

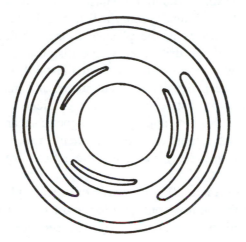

Fig. 10.12: Cross sectional view of magnetic islands formed by tearing modes at flux surfaces with q being a rational number. Where q ≠ rational, a sequence of MHD-equilibrium flux surfaces appears.

(c) Microinstabilities

Microinstabilities are often associated with non-Maxwellian velocity distributions. The deviation from thermodynamic equilibrium means that there is free energy which can drive instabilities, often evolving into plasma turbulence. Also, nonuniformity and anisotropy of distributions can give rise to instabilities. Hence it is the particle kinetic effects that play an important role here, and the plasma cannot be expected to behave as a simple fluid and therefore cannot be treated as such anymore. It rather requires a kinetic description.

The electron velocity distribution becomes increasingly anisotropic as the plasma density decreases since, in order to carry a given plasma current, the individual electrons are further required to align their velocities with the current density direction. This drift velocity tends to make the velocity distribution function more and more asymmetric and hence unstable. On the other hand, the electrons will encounter collisions which randomize their velocities and thus reestablishes symmetry in the velocity distribution. However, as the density is further decreased, this stabilizing effect due to collisions is ultimately overcome by the destabilizing effect of the increasing drift velocity.

Further, anisotropy occurs in a plasma when it is confined by mirror fields, since particles having a large $v_{\parallel}$-component will escape and are therefore lost from the distribution. Hence, also the trapped particles in a tokamak, which bounce back and forth in the local mirror fields, can constitute a source of instabilities, preferably when the perturbation frequency is less than the bounce frequency. These then are classified as trapped particle instabilities and are still being investigated for effects on increased cross-field diffusion in tokamaks.

As a current is driven through the plasma or a beam of high energetic particles is injected, the different species will drift relative to one another. The drift energy can excite waves in the plasma. Since oscillation energy may be gained at the expense of the drift energy, the disturbance can grow. Such an instability is called a 'two-stream' or beam-plasma instability.

Where there is a steep density gradient in the plasma, an instability may appear due to the electron drift caused by the gradient and is called the drift instability. It can be stabilized by appropriate magnetic shear.

Microinstabilities are a large and complex field in plasma physics, which is still under investigation and many theoretical predictions are still to be proven by experiments.

Generally, there are three effective ways to prevent plasma instabilities: (i) magnetic shear, (ii) minimum-**B** configuration and (iii) dynamic stabilization by oscillating **E**- or **B**-fields or by proper-phase force feedback.

10.6   Stellarator Concept

As discussed in Sec. 10.1, the confinement of a plasma in a toroidal magnetic field requires a rotational transform of field lines in order to prevent local charge concentrations, plasma polarization and drifts to the wall. Unlike the tokamak which carries an externally induced current in its plasma, stellarator devices do not. They also feature a toroidal geometry but render the confining magnetic field lines helical, i.e. compel a rotational transform, either by a deformation of the torus itself–such early concepts showed poor stability–or by helical or contorted coil currents external to the plasma. These currents, as shown in Fig. 10.13, pass through helical conductors winding around the torus and make the magnetic field lines take on the form of a spiral. A rotational transform produced in this way, as well as the closed magnetic flux surfaces thus rendered, exist in a vacuum field and do not rely on a plasma current induced in a pulsed manner. In stellarators, all magnetic fields providing confinement of the plasma are generated by means of currents flowing in external conductors. Hence, as an important advantage, stellarators allow for steady-state confinement and continuous fusion reactor operation.

In practical terms , a helical winding is a loosely wrapped solenoidal winding and generates, as desired, a toroidal and poloidal field. Furthermore, if viewed from above the torus, it represents also as a loosely wrapped vertical field coil and hence generates a vertical field as well. To eliminate this contribution, currents in adjacent helical windings of the same pitch flow in opposite directions canceling out one another's vertical fields and also their toroidal fields, on average. Thus, as seen in Fig. 10.13, a separate set of coils is needed to provide the essential toroidal magnetic field . Therefore, a stellarator still requires toroidal field coils as shown. The poloidal field produced from the helical windings together with the toroidal field from the separate toroidal field coils result in a flux which twists the magnetic field lines as they pass around the torus and thus generate magnetic surfaces of the shape shown in Fig. 10.14. Evidently, the geometrical simplicity of axisymmetry is lost. It is noted that the establishment of closed magnetic surfaces is possible only in a restricted region of the minor cross section of the toroidal tube. For a stellarator with $\ell = 3$ pairs of helical coils of opposite currents, we illustrate in Fig. 10.15 the shape of the generated magnetic surfaces. Closed magnetic surfaces are observed within the cross-sectional area embraced by the dashed separatrix line which may be identified also as the last closed magnetic surface. Outside this separatrix the field lines wrap around the individual conductors.

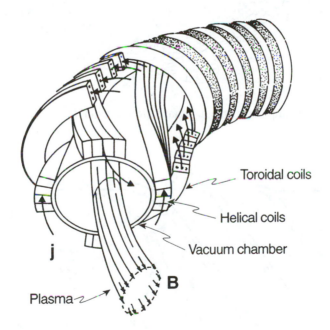

Fig. 10.13: Stellarator magnetic field configuration generated by external helical currents of opposite direction in alternate coils and, additionally, by toroidal field coils.

Due to the absence of a current in a stellarator, Ampere's law (compare with Eq. (9.48)) yields here

$$\oint_s \frac{\mathbf{B}}{\mu_o} \cdot d\mathbf{s} = 0 , \tag{10.51}$$

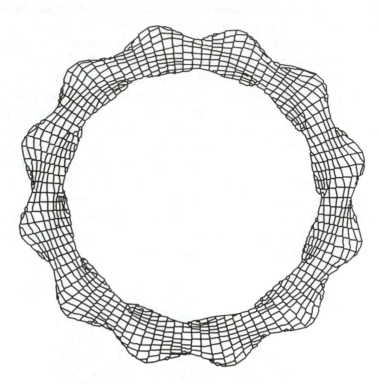

Fig. 10.14: Complete magnetic surface viewed from the top of the stellarator.

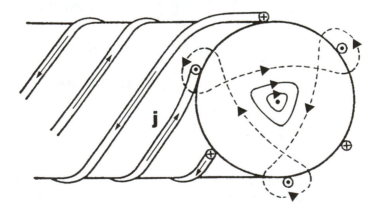

Fig. 10.15: Cross section of nested magnetic surfaces generated by an $\ell = 3$ stellarator.

meaning that the line integral of the poloidal component B_θ of the magnetic field vanishes along a contour s encircling the magnetic axis on each magnetic flux surface. For this to be true, the poloidal field must change sign and magnitude along s. Hence, unlike in a tokamak, the magnetic field lines in a stellarator do not wrap monotonically around the toroidal tube. Rather, they appear to oscillate periodically according to the qualitative structure given in Fig. 10.16a. Each such so-called fundamental field period incrementally rotates the field lines in the poloidal direction. Yet another inhomogeneity of the magnetic field is to be considered, which is due to the curvature associated with torus geometry. The resulting variation of the magnetic field along the toroidal direction is illustrated in Fig. 10.16b where the deep and more frequent oscillations of B are caused by the helical windings alternately carrying currents of different direction, and where the slow modulation of B corresponds to the toroidal curvature. It is obvious that in addition to the magnetic mirrors in the toroidal field which lead to particle trapping as in Sec. 10.3, there are also local mirrors of the helical field.

Analyzing the particle motion in such magnetic field configurations yields three distinctive types of orbits: (i) circulating particles which pass entirely around the torus without encountering a reflection, (ii) so-called 'helically trapped' particles reflected in the local mirrors of the helical field, and (iii) 'toroidally trapped' particles tracing banana orbits as they are reflected in the toroidal magnetic mirrors known from Sec. 10.3. It is possible that a helically trapped particle appears to be toroidally trapped as well. Such a particle is then called a 'superbanana particle'.

Another approach to establish a rotational transform for toroidal plasma confinement is to partially rotate non-circular toroidal field coils, one with respect to the other. Further, specific designs–with their general principle demonstrated in Fig. 10.17–allow for practical modular composition. For reasons of simplification, in Fig. 10.17 the helical windings of a stellarator are reduced to consist of only two conductors with currents of opposite sign. Such an $\ell = 1$ configuration permits replacement by modules which combine parts of the helical coils with additional meridian-ring conductors. Advanced modular designs utilize non-planar twisted coils, that are sophisticated spatial elements as illustrated in Fig. 10.18. Modular coils are favourable from an engineering standpoint of view because they allow for assembly and disassembly of the coils without having to unwind or disconnect helical conductors encircling the major torus axis.

Closely related to the stellarator configuration is the so-called 'torsatron' in which the rotational transform is produced again by helical windings, however with like current directions. In this design there is no need for toroidal field coils, but, instead, equatorial ring conductors may be applied to generate a transverse field which can compensate the vertical magnetic field produced by the torsatron as a result of like helical current directions. This necessity of a compensating vertical field may be avoided by specifically winding the helical conductors according to

$$m\theta = \phi + \alpha \sin \phi + \beta \sin(2\phi) \qquad (10.52)$$

with m being an integer and α, β denoting chosen constants. Such a device is called the 'ultimate' torsatron.

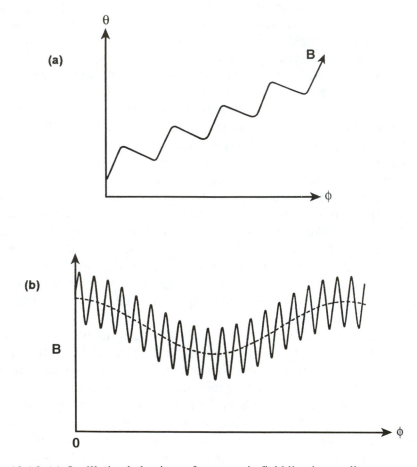

Fig. 10.16: (a) Oscillative behaviour of a magnetic field line in a stellarator on a flux surface as a function of the poloidal (θ) and toroidal (ϕ) angle coordinate. (b) Varying magnitude of the magnetic field in a stellarator configuration as a magnetic field line is followed around the torus.

Much theoretical and experimental investigation is still devoted to the determination of the maximum plasma-beta in a stellarator/torsatron for which stable equilibrium can be sustained. It is thought that β's of several percent, perhaps up to 10%, can be stably achieved if the configuration exhibits a helical magnetic axis.

Non-axisymmetric configurations such as the stellarator may lead–in comparison with axisymmetric devices–to more complex transport processes due to the greater variety of particle orbits. The quality of particle and energy confinement is dominantly determined by the trapping of particles in the various ripples of the magnetic field and by the frequency of collisions occurring in the plasma. Collisions may scatter particles from one region of trapping to an adjacent region and thereby alter the type of trapping. Interestingly, in non-axisymmetric toroidal systems, the electron and ion components of the confined plasma diffuse independently of each other.

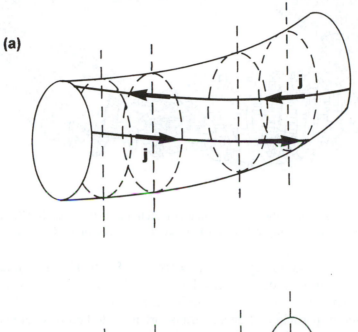

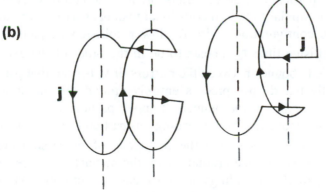

Fig. 10.17: Depiction of replacement of helical conductors by modular elements: (a) helical stellarator windings; (b) modular coils generating a stellarator magnetic configuration much the same as established by (a).

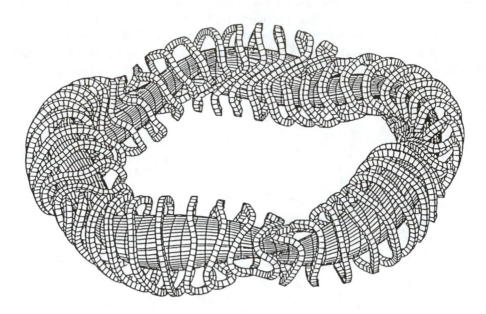

Fig. 10.18: Design of the advanced modular coil stellarator Wendelstein VII-X (with
permission of G. Grieger, MPI for Plasma Physics, Garching, Germany).

For very high collision frequencies $v_c = 1/\tau_c$ (see Eq. (6.17)) in a stellarator or
a torsatron, the particles do not travel sufficiently long distances without a
scattering encounter to be reflected by either the helical or the toroidal mirrors.
As a consequence, particle and energy transport in a stellarator operated in this
collisional regime is similar to that in collisional tokamaks. However, differences
between these two configurations arise for lower collision frequencies, that is
when the mean collision time τ_c appears to be in the order of the average time
needed by a particle to bounce between the mirrors. If v_c is such that particles can
be trapped helically but do not precess entirely around the minor torus axis
before undergoing a collision, the helically trapped particle will drift from one
magnetic surface to another. This spreading of particles obviously enhances
particle diffusion. For v_c lower than the frequency of precession around the
magnetic axis, i.e. the single line around which the magnetic surfaces appear to
be nested, these particle spreadings tend to cancel out, and the diffusion
coefficient in this regime is expected to decline with decreasing collision
frequency. If, however, superbananas are present, as featured by stellarators, the
diffusion coefficient will not decline immediately with a reduced v_c, but rather it
remains constant at its high value over a limited collision frequency interval and
thus exhibits the so-called superbanana plateau. Finally, as v_c becomes smaller
than the superbanana bounce frequency, the diffusion coefficient is observed to

decrease in stellarators as well.

Thermal energy diffusion appears to follow a dependence on v_c similar to that seen for particle diffusion. The plasma-energy-confinement time of stellarator/torsatron devices is thought to scale similarly to that found for tokamaks, except for the weak-collision regime where the spreading of helically trapped particles enhances the diffusion. Suppressing this contribution is the objective of advanced stellarator designs.

In conclusion, we summarize some potential advantages of the stellarator/torsatron concept: Steady-state magnetic fields simplify the magnet design. Unlike in tokamak reactors, there is no need for pulsed superconducting coils and corresponding energy storage to drive these pulsed coils. Since a toroidal plasma current is not needed, a potential source of instabilities is eliminated. Early predictions of enhanced transport losses and increased instability have not materialized. Stellarator and torsatrons appear to be operating as effective plasma confinement machines, with dimension and performance parameters comparable to those of similar toroidal magnetic devices. The high aspect ratio, the absence of transformer coils and, particularily, modular construction make stellarator/torsatron devices well accessible. Further, steady-state operation of an ignited plasma would allow for a simplified blanket design due to reduced material durability requirements.

10.7 Alternate Closed Configurations

While alternate magnetic concepts may differ from tokamaks in geometry, size, time scales, input power requirements and technology, the principal objectives remain, that is the heating of a D-T plasma to fusion ignition and confining it sufficiently long to yield a net energy gain.

Out of many different designs proposed and discussed in the literature, one such concept is the so-called 'bumpy torus', which links a number of mirror sections end to end into a high-aspect-ratio torus, depicted in Fig. 10.19. As illustrated, several mirror coils are equally spaced in a toroidal array. The plasma contained by this magnetic configuration threads the bores of these axisymmetric coils and thus takes on the shape of a bumpy toroidal 'sausage'. The magnetic field lines close on themselves and the plasma particles are confined in two ways: trapped particles reflect back and forth in individual mirror sections, and passing particles circulate around the major circumference of the bumpy torus plasma. To stabilize such a configuration, electron cyclotron resonance heating (ECRH) is applied to form an annular high-energy electron plasma in the central part of each mirror section. When the currents generated by these hot-electron rings are sufficient to provide a minimum-B configuration, they thus stabilize the toroidal core plasma. In such a reactor, which is called the Elmo (<u>e</u>lectrons with <u>l</u>arge <u>m</u>agnetic <u>o</u>rbits) Bumpy Torus, the hot-electron annuli would typically possess densities of about 10^{18} m^{-3} and temperatures T$_e$>100 keV, while the core plasma exhibits a density in the order of 10^{20} m^{-3} and a temperature of about15 keV. It is thus possible that the beta-value of the confined bumpy torus plasma can become comparable to that of the annuli. The toroidal plasma appears to be macroscopically stable as long as its beta-value is smaller than, or at most, approximately equal to the beta of the annuli. To produce the stabilizing minimum-B property, the latter beta has to exceed a threshold value in the range 5-15% depending on the annular shape. It was experimentally shown that such hot-electron rings can produce beta's up to 50% in steady-state operation. Hence, the core plasma-β of an Elmo Bumpy Torus may also be established at substantially increased values in comparison to the tokamak. As a consequence, the fusion power density, as limited by the magnetic pressure, Eq. (4.14), is elevated or, for a given power density, the magnetic field requirements are significantly reduced. Another advantage over the tokamak is the large aspect ratio ($\approx$ 5-10 times greater) allowing for simpler engineering design and construction. Further, there is no need for power interruption as associated with pulsed operation; an Elmo Bumpy Torus can thus be operated in a steady-state mode.

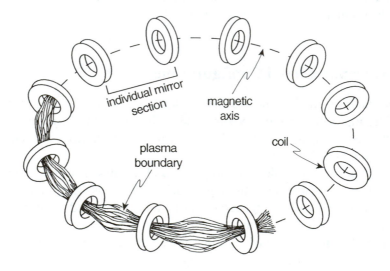

Fig. 10.19: Toroidal plasma confined by toroidally linked magnetic mirror coils constituting the bumpy torus concept.

Avoiding the production of hot-electron annuli, which is relatively inefficient by means of ECRH and leads to increased radiation losses, a toroidal minimum-B configuration can also be generated by toroidally linking modular coils of specific shape such that each already represents a minimum-B magnetic mirror. To introduce a rotational transform, the coils, which do not exhibit poloidal symmetry, are rotated about the magnetic axis with respect to each adjacent coil.

A device combining the effect of a z-pinch with that of a θ-pinch–recall Fig. 9.10–will contain a plasma with currents in the axial as well as in the poloidal direction and thus generate a confining magnetic field consisting of helical field lines. Due to the form of the field lines, this configuration is called a screw-pinch. Though similar to the tokamak, it is operated at relatively high β≈20%, but features only very short periods of sufficient plasma confinement.

Another toroidal confinement concept, which has received great attention due to its improved stability against MHD modes, is the so-called Reversed Field Pinch (RFP). It is much like the tokamak: the plasma is confined by a combination of toroidal and poloidal magnetic fields with the latter generated by a toroidal plasma current induced by transformer action. The toroidal field $\mathbf{B}_\phi$ is established primarily by external coils. The essential difference, however, is that in the RFP the plasma currents parallel to the toroidal minor axis do not only produce the poloidal field, $\mathbf{B}_\theta$, but also diamagnetically alter the toroidal field such that $\mathbf{B}_\phi$ can change sign near the plasma boundary (field reversal). Further, the plasma current and B_θ in RFP's are much stronger than in comparable tokamaks, whereas B_ϕ is modest. This gives rise to strongly sheared magnetic field lines with their pitch increasing rapidly with greater radial distance. The Reversed Field Pinch configuration is produced by the high magnetic shear near the edge of the plasma which suppresses local MHD instabilities. The field reversal is suggested to emerge from a turbulent state as a self-organization mechanism.

In contrast to a tokamak, where the safety factor q has to meet the Kruskal-Shafranov stability criterion, q(r)>1 everywhere and q(r=a) ≥ 2.5 (see Sec. 10.1), an RFP features q(r)<1 with a negative q(r→a) consistent with the reversal of the toroidal field component in this edge region. The evolution from a tokamak plasma to an RFP requires the presence of an electrically-conducting shell just outside the toroidal plasma or of closely fitting external conductors for assisting the tokamak plasma to remain confined while reducing q and turning to the RFP configuration. MHD stability theory for RFP indicates the plasma-β limitation at the high value of ~30%. Due to the high β, the deployment of advanced fusion fuel cycles in these devices is conceivable. A fusion plasma system not constrained by the Kruskal-Shafranov criterion provides the profit that it can be heated ohmically to ignition, if the energy confinement is good. Experiments, however, have shown so far a τ_E lower than that for tokamaks of similar size. An obvious advantage over the tokamak is the elimination of the requirement of minimizing the aspect ratio, such as previously demanded by Eq. (10.49). Hence, simplified designs with good maintenance access are possible. A handicap common with tokamaks is that the RFP is a pulsed device as well.

The reactor concepts discussed so far are physically large, they employ complex technology, represent expensive designs, and possess only a relatively low power density. Evidently, a high power density would be a desirable feature of a fusion reactor. This may be accomplished in compact power reactors which achieve the same total power as a conventional magnetic fusion device in a significantly smaller geometry. Among various designs proposed in this context, e.g. a compact RFP, we choose here to describe the spheromak reactor as a distinctive representative.

A spheromak is an advanced toroidal plasma containment device in which the confining magnetic configuration, as displayed in Fig. 10.20, is characterized by an extremely low aspect ratio and by the absence of external toroidal fields. With the minor plasma radius a ~ R_o, this configuration appears almost like a sphere and, obviously, this has inspired the given name. An axial current flows through a field-reversed θ-pinch plasma to internally produce the toroidal field. Thus, both the poloidal and the toroidal magnetic field are self-generated. Only the steady magnetic-bottle field is provided externally by coils. Topologically, spheromaks are open confinement systems with the plasma, however, contained within a closed separatrix surface. The region inside this separatrix is–similar to the RFP– associated with q<1, while, at the plasma boundary, the safety factor is zero. The great advantage of this design is its geometric simplicity and compactness. Experiments with spheromak configurations to date could be operated for only short pulse lengths.

Most pulsed toroidal confinement systems suffer from extremely low periods during which the plasma can be stably contained, and from relatively large radiation power losses due to the impurities released by intense plasma wall interaction. Though there exists a number of confinement concepts which are not discussed here but nevertheless are interesting in their specific design and/or the physics to be applied, we conclude this chapter by recalling that–among the closed magnetic systems–tokamak and stellarator concepts are the most promising candidates to be utilized for future fusion reactor operation.

Problems

10.1   Find the drift velocity, as given in Eq. (10.5) for the case of a purely toroidal magnetic field, by combining the effects of the grad-B drift and the curvature drift, which are simultaneously present in a curved B-field, for the

specific magnetic field $\mathbf{B} = \left(B_r = 0 \, , \; B_\theta = \dfrac{r}{a} B_\theta(a) \, , \; B_\phi = \dfrac{R_o}{R_o + r \cos\theta} B_\phi(R_o) \right)$

10.2   Assuming that a scaling law of the form $N\tau_{E^*} \approx 1.7 \times 10^{-19} \left(\dfrac{a}{R} \right)^2 N^2$

may be applied to a tokamak reactor, estimate the size of such a reactor having a

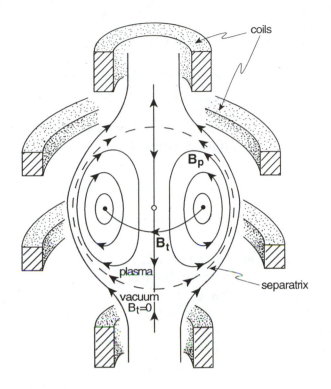

Fig. 10.20: Poloidal and toroidal magnetic fields, $\mathbf{B_p}$ and $\mathbf{B_t}$, specific to a spheromak configuration. Beyond the separatrix (dashed line) $B_t=0$.

circular plasma cross-section if it is to achieve ignition and contains a 50:50% d-t fusion plasma with an average ion density $N_i = 10^{20}$ m^{-3} at average temperature $T_i=T_e=30$ keV producing 1000 MW of fusion power. Use the ignition criterion of Sec. 8.4.

10.3   Using $\mathbf{j} = \nabla\times\mathbf{B}$, show that for a tokamak the magnetic field is proportional to 1/R.

10.4   What perspectives do high-beta devices offer relative to low-beta tokamaks? How can a high-beta configuration be realized?

10.5   For each of the following open magnetic confinement fusion reactor concepts, draw a sketch of the concept, label the major components, and describe their purpose. Describe how fusion fuel ions are confined, and list what the energy and particle losses are in the system, and where they occur. Describe all the different electrical currents , magnetic fields, and their purpose.
    (a) Tokamaks

   (b) Stellarators
   (c) Spheromaks
   (d) Reversed Field Pinch (it is a toroidal system)

10.6   List the favourable characteristics desirable for a future fusion power reactor.

10.7   Design an axisymmetric d-t tokamak reactor with circular cross-section capable of fusion plasma ignition demonstration assuming the empirical energy confinement time scaling

$$\tau_E[s] = 0.00338 \left(I_p[\text{MA}]\right)^{0.85} \left(a[\text{m}]\right)^{0.3} \left(R_o[\text{m}]\right)^{1.2} \left(N_e[\text{m}^{-3}]\right)^{0.1} \left(B_t[\text{Tesla}]\right)^{0.2} \left(f_{c,dt} P_{fu}^*[\text{MW}]\right)^{-0.5}$$

where P_{fu}^* denotes the total fusion power in the plasma volume, and taking the following fixed parameters:
   ·   ratio of first wall to plasma radius $r_w/a = 1.25$
   ·   50:50% deuterium-tritium fuel mixture
   ·   electron density $N_e = 0.8 \times 10^{20}$ m^{-3}
   ·   equal ion and electron temperature, $T_i = T_e = 20$ keV
   ·   plasma current $I_p = 18$ MA
   ·   toroidal magnetic field $B_t = 6$ Tesla
   ·   cyclotron radiation loss parameter $\psi = 0.001$
Specify the relevant plasma and reactor parameters allowing for ignition (i.e. so that Eq. (8.30) is met) and consistent with the constraints and requirements discussed in Sec. 10.4, in particular Eqs. (10.40b), (10.42) and (10.46), as well as with the engineering constraint of limiting the thermal power flux through the first wall by $P_w \leq 1.5$ MW m^{-2}.

11.   Inertial Confinement Fusion

An additional approach to confining fusion reactants, completely distinct from that of magnetically confined systems, is inertial confinement fusion which involves compressing a small fuel pellet to very high density by an intense pulse of energy. This compressive pulse of energy may be supplied by lasers or ion beams. We now investigate several issues fundamental to this approach to fusion.

11.1   Basic Concepts

Both magnetic and inertial confinement fusion involve two key processes for the attainment of a viable fusion energy system:
1. heating and ionization of the fuel to high temperature to achieve a favourable fusion reaction rate density, and
2. confinement of the fuel for a sufficiently long time to yield a net energy gain.

The particle density in a magnetically confined plasma is expected to not significantly exceed 10^{21} m^{-3}, which constitutes a very low density gas when compared to atmospheric particle densities of about 10^{25} m^{-3} at standard temperature and pressure. In contrast, inertial confinement typically requires fuels compressed to densities that are several orders of magnitude greater than solid–approaching number densities of 10^{31} to 10^{32} m^{-3}–exceeding even densities found in stars. A major feature of this high density is that since the fusion reaction rate is proportional to the density squared, the inertial confinement time required for a net energy gain will be significantly smaller than those in magnetic confinement devices. These points can be put into perspective by comparing the Lawson criterion-like requirements for inertial and magnetic confinement. For magnetic confinement, magnetic field limitations typically restrict ion densities to the order of 10^{21} m^{-3} with a confinement time necessary for energy break-even of about 1 s. On the other hand, for inertial confinement, the compressed density can be 10^{31} m^{-3} over a time interval of the order of 10^{-9} s.

Interest in inertial confinement fusion energy emerged later than that in magnetic confinement fusion. Its relevance to power production became apparent when it was recognized that concentrated beams from powerful pulsed lasers could be used to initiate a compressive process in a small solid or liquid target pellet, possibly resulting in a sufficient number of fusion reactions to yield a net energy gain.

The sequence of events in inertial confinement fusion can be briefly described as follows. A small pellet, with a radius less than ~5 mm and containing a mixture of fuel atoms, is symmetrically struck by energetic pulses of electromagnetic radiation from laser beams or by high energy ion beams from an accelerator, Fig. 11.1a. Absorption of this energy below the surface of the pellet leads to local ionization and a plasma-corona formation, Fig. 11.1b. The important consequences of these processes are an outward directed mass transfer by ablation and–by a rocket-type reaction–an inward directed pressure-shock wave leading to compressing and heating of the target, Fig. 11.1c. A follow-up shock wave driven by the next laser or ion beam pulse will then propagate into an already compressed region where it travels faster than its predecessor. Subsequent shock waves can thus propagate even more quickly. Tuning the beam's pulse repetition rate such that the shock waves arrive at the pellet's core simultaneously will provide for an adequately compressed state possessing temperatures suitable for initiating a substantial fusion burn. With the temperature and fuel density sufficiently high, the fusion reactions will occur until the pellet disassembles in a micro-explosion due to its excessive energy content, Fig.11.1d. The disassembly typically takes place in a time interval of about 10^{-8} s, corresponding to the propagation of a pressure wave across the pellet with sonic speed v_s.

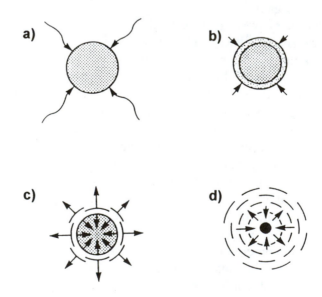

Fig. 11.1: Stages associated with inertial confinement fusion: (a) irradiation with high intensity beams, (b) corona formation, (c) ablation and compression, (d) heating, fusion, and disassembly.

Experience with inertial confinement has shown the importance of several processes and phenomena. For example, it is essential that the incident laser or ion beam strike the pellet symmetrically and that efficient energy coupling between the beam and the target be attained. The inner core should reach a high density very quickly before thermal conductivity heats the central region causing an internal pressure build-up that opposes high compression. A substantial fraction of the nuclear fuel should also burn before pellet disintegration.

We note that laser beams can penetrate to deeper layers of the pellet when they possess a higher frequency. Hence, intensive short-wavelength lasers are sought as drivers for inertial confinement fusion, evidently with a reasonable efficiency also required. A phenomenon of concern is that very high energy electrons generated in the initial laser light absorption process will penetrate into the centre of the target before the arrival of the dominant pressure wave thereby causing an undesirable preheating of the central core region resulting in an outward force effect to retard compression. Accelerators, on the other hand, transfer the beam energy more directly to ions in the target and can therefore be significantly more efficient; this provides some appeal for the use of light or heavy ion accelerators for such purposes.

11.2   Rho-R Parameter

Some useful parameter estimates about inertial confinement fusion can be obtained by an analysis of selected particle kinetics and energy transfer processes. Consider, therefore, a spherically symmetric pressure wave converging towards the pellet centre. Suppose that ignition and burn conditions are attained when the radius of the compressed pellet is R_b and hold over a burn time τ_b, during which the pressure generated by the shock waves and the heating due to fusion reactions causes the pellet to expand to an extent where the density, and hence the fusion reaction rate, have decreased to insignificant levels.

At any time during this nuclear burn period, we take the total number of ions in the burning part of the pellet to be $N_b{}^*$. This total ion population decreases with time because of fusion reactions which occur at the rate

$$\frac{dN_b^*}{dt} = -2\int_{V_b} R_{fu} d^3r. \tag{11.1}$$

Here R_{fu} is the fusion reaction rate density with each fusion event destroying two ions; the integration is over the pellet burn volume V_b. Substituting for the fusion reaction rate density involving deuterium and tritium gives

$$\frac{dN_b^*}{dt} = -2 \int_{V_b} N_d(t) \, N_t(t) <\sigma v>_{dt} d^3r$$

$$= -2 \int_{V_b} \left(\frac{N_i(t)}{2} \right) \left(\frac{N_i(t)}{2} \right) <\sigma v>_{dt} d^3r \qquad (11.2)$$

$$= -\frac{1}{2} \int_{V_b} N_i^2(t) <\sigma v>_{dt} d^3r \, ,$$

where a 50:50% tritium-deuterium ion density composition is assumed.

To roughly assess the requirements for viable fusion burn, allow us now to consider the somewhat unsound assumption that during the burn time τ_b, the fuel ion density and temperature are a function of time only and uniform distributions exist in the burning part of the pellet. Evidently then, the total number of fuel ions $N_b^*(t)$ and the fuel ion density $N_i(t)$ in the burn volume are related by

$$N_i(t) = \frac{N_b^*(t)}{V_b} \, , \qquad (11.3)$$

and therefore Eq.(11.2) gives

$$V_b \frac{dN_i}{dt} = -\frac{N_i^2(t)}{2} <\sigma v>_{dt} \int_{V_b} d^3r = -\frac{N_i^2(t)}{2} <\sigma v>_{dt} V_b \, . \qquad (11.4)$$

The variables can now be separated and integrated to give

$$\int_{N_{i,o}}^{N_{i,f}} \frac{dN_i}{N_i^2(t)} = -\frac{1}{2} \int_0^{\tau_b} <\sigma v>_{dt} (t) \, dt \, . \qquad (11.5)$$

Here $N_{i,o}$ is the ion density at the beginning of the burn, $t = 0$, and $N_{i,f}$ is the fuel ion density at the end of the burn, $t = \tau_b$. Integration and rearrangement of the terms yields for the burn time

$$\tau_b = \frac{2}{<\sigma v>_{dt}} \left(\frac{1}{N_{i,f}} - \frac{1}{N_{i,o}} \right), \qquad (11.6)$$

where $\overline{<\sigma v>_{dt}}$ denotes the fusion reactivity parameter averaged over the burn period according to

$$\overline{<\sigma v>_{dt}} = \frac{1}{\tau_b} \int_0^{\tau_b} <\sigma v>_{dt} (t) \, dt \, . \qquad (11.7)$$

It will be useful to introduce the symbol f_b for the fraction of fuel burned during τ_b,

$$f_b = \frac{N_{i,o} - N_{i,f}}{N_{i,o}} \, . \qquad (11.8a)$$

Thus $\qquad\qquad\qquad N_{i,f} = N_{i,o}(1 - f_b) \qquad\qquad\qquad (11.8b)$

and we also use

$$\rho_b \approx N_{i,o} \overline{m_i} \qquad (11.9)$$

where ρ_b is the pellet density during the burn, $\overline{m_i}$ is the average ion mass, and the mass contributions of the electrons have been neglected, attributable to m_e being three orders of magnitude smaller than $\overline{m_i}$. Substitution then yields the explicit expression for τ_b, Eq.(11.6), as

$$\tau_b = \frac{2}{<\sigma v>_{dt}} \frac{\overline{m_i}}{\rho_b} \left(\frac{f_b}{1-f_b} \right), \qquad (11.10)$$

with ρ_b still unknown because $N_{i,o}$ is not known.

The fusion burn will continue over the time interval of pellet disassembly given by

$$\tau_{dis} = \frac{R_b}{v_{dis}}, \qquad (11.11)$$

where v_{dis} is the speed at which the core mass moves outward. The corresponding kinetic energy is of the order of the ion thermal energy, Eq. (2.19c), so that we may use

$$\tfrac{1}{2} \overline{m_i} v_{dis}^2 \approx \tfrac{3}{2} kT_i, \qquad (11.12)$$

and therefore we derive here

$$v_{dis} \approx \sqrt{\frac{3kT_i}{\overline{m_i}}}, \qquad (11.13)$$

which can be compared with a previous assessment, Eq.(4.4).

With the burn time evidently not exceeding the disassembly time, i.e. $\tau_b < \tau_{dis}$, we require Eqs.(11.10) and (11.11) to satisfy

$$\frac{2}{<\sigma v>_{dt}} \frac{\overline{m_i}}{\rho_b} \left(\frac{f_b}{1-f_b} \right) < \frac{R_b}{v_{dis}}. \qquad (11.14)$$

Rearranging this expression yields the important Rho-R parameter for inertial confinement fusion:

$$\rho_b R_b > \frac{2\overline{m_i} v_{dis}}{<\sigma v>_{dt}} \left(\frac{f_b}{1-f_b} \right). \qquad (11.15)$$

This specifies the conditions on the pellet density and pellet radius at the beginning of the fusion burn that are required for a specified burn fraction with the reaction occurring at some average temperature. A useful numerical value for this parameter is obtained by taking $\overline{<\sigma v>}_{dt} \approx <\sigma v>_{dt}$ (T = 20 keV) and an estimate for v_{dis} also at this temperature, yielding for a 50% burn fraction

$$\rho_b R_b > 3 \ g \cdot cm^{-2}. \qquad (11.16)$$

For $R_b \approx 1$ mm this demands a density $\rho_b \approx 30$ g·cm^{-3} and is indeed very high compared to d-t liquid density of $\rho_\ell \approx 0.2$ g·cm^{-3}. Thus, the compression of the initial fuel pellet by a factor of about 10^3 to 10^4–relative to liquid density–appears to be necessary for a satisfactory burn.

11.3   Energy Balance

The Lawson criterion and general energy flow analyses of Ch. 8 can be applied to inertial confinement as well as to magnetic confinement. However, the parameters of interest and the nomenclature is different for each. We now re-examine the energy balance specifically for an inertial confinement fusion system, where one pulse is still taken as the characteristic time interval for which an energy balance is established.

For present purposes, we define the following three energy components necessary in a parametric analysis of an inertial confinement fusion system, Fig. 11.2:

$E_{be}^* = $ energy contained in the laser or ion beam which triggers compression;

$E_{th}^* = $ thermal energy of the compressed target ions and electrons following impingement of the beam;

$E_{fu}^* = $ fusion energy released during the associated burn time τ_b.

Not all of the beam energy will appear as thermal energy of the ions and electrons in the target; a fraction may be reflected or scattered and some energy is carried off with the ablated outer layer. Hence, a coupling efficiency η_c can be defined which relates E_{be}^* and E_{th}^* by

$$E_{th}^* = \eta_c E_{be}^* , \quad 0 < \eta_c < 1 . \tag{11.17}$$

A characteristic pellet energy multiplication M_p relates E_{fu}^* to E_{be}^* by

$$E_{fu}^* = M_p E_{be}^* , \tag{11.18}$$

and for an energetically viable system, M_p has to substantially exceed 1. Note that η_c and M_p are design parameters of the system.

The overall energy flow for an electricity producing inertial confinement fusion reactor system is suggested in Fig.11.2 for which the station electrical energy output is given by

$$E_{net}^* = \eta_{fu} E_{fu}^* - E_{in}^* . \tag{11.19}$$

Here η_{fu} is the efficiency of converting the fusion energy into electrical form and E_{in}^* is the circulating electrical energy component required to sustain the lasers or ion accelerators. The conversion of this electrical energy into beam energy is taken to occur with an efficiency η_{in} defined by

$$\eta_{in} = \frac{E_{be}^*}{E_{in}^*}, \quad 0 < \eta_{in} < 1 . \tag{11.20}$$

The station electrical energy output can be compactly written by defining an electrical energy multiplication as

$$M_e = \frac{\eta_{fu} E_{fu}^*}{E_{in}^*} , \tag{11.21}$$

so that the station energy production, Eq.(11.19) becomes

$$E_{net}^* = \eta_{fu} E_{fu}^* \left(1 - \frac{1}{M_e} \right) . \tag{11.22}$$

The essential requirement for a viable inertial confinement fusion system is therefore

$$M_e > 1 . \tag{11.23}$$

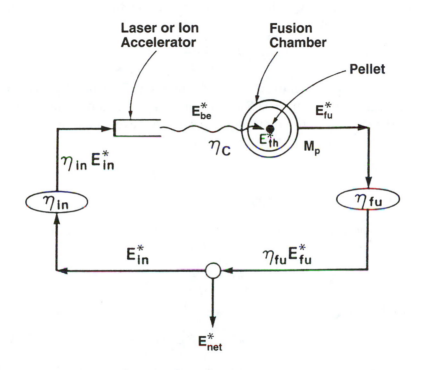

Fig. 11.2: Energy components for an inertially confined fusion system applicable to a compression pulse.

This energy viability criteria can be related to the several conversion efficiencies and the pellet multiplication already defined. A substitution of Eqs.(11.18), (11.20) and (11.21) into Eq.(11.23) yields

$$\eta_{in} \eta_{fu} M_p > 1 , \tag{11.24}$$

and thus specifies the necessary pellet energy multiplication required. As currently envisioned, lasers are relatively inefficient with $\eta_{in} \sim 0.06$, while $\eta_{fu} \sim$ 1/3 for conventional energy conversion; this yields a requirement of $M_p > 50$. This demanding result can be reduced if the driver is more efficient; for example for ion accelerators, $\eta_{in} \sim 0.3$ may be possible, giving $M_p > 10$.

Further, the beam coupling efficiency enters via

$$M_p = \frac{E_{fu}^*}{E_{be}^*} = \eta_c \frac{E_{fu}^*}{E_{th}^*} \tag{11.25}$$

and upon substitution in Eq.(11.24) gives

$$\eta_{in} \eta_c \eta_{fu} \left(\frac{E_{fu}^*}{E_{th}^*} \right) > 1 . \tag{11.26}$$

Note that (E_{fu}^* / E_{th}^*) is the ratio of fusion energy produced to the energy deposited in the pellet and hence, in analogy to Eq. (8.6), can be identified as the according pellet plasma Q-value which, upon introduction in Eq. (11.26), has to satisfy the requirement

$$Q_{pp} > \frac{1}{\eta_{in} \eta_c \eta_{fu}} \tag{11.27}$$

for energy viability. Evidently, a very high coupling efficiency η_c is desired. For example, for a laser with $\eta_c = 0.05$ and the previous η_{in} and η_{fu} values, we require $Q_{pp} > 1000$. One way to meet this requirement is to have a high fusion gain pellet which in turn implies a very high compression.

11.4   Compression Energy

Estimates of the beam energy required to compress a pellet to burn conditions can be obtained by starting with Eq.(11.17) in the form

$$E_{be}^* \equiv \frac{E_{th}^*}{\eta_c} \tag{11.28}$$

and hence, from Eq.(11.18)

$$E_{th}^* = \frac{\eta_c}{M_p} E_{fu}^* . \tag{11.29}$$

Based on our previous considerations, the total thermal energy in the pellet of radius R_b at the beginning of the fusion burn is explicitly given by

$$E_{th}^*(0) = \left(\tfrac{3}{2} N_{e,o} kT_e(0) + \tfrac{3}{2} N_{i,o} kT_i(0) \right) \left(\tfrac{4}{3} \pi R_b^3 \right) = 4\pi N_{i,o} kT_{i,o} R_b^3 \tag{11.30}$$

where we have taken

$$N_{e,o} = N_{i,o} \quad \text{and} \quad T_e(0) = T_i(0) = T_{i,o} . \tag{11.31}$$

According to Eq. (11.28), the beam energy requirement is therefore

$$E_{be}^* = \frac{1}{\eta_c} \left(4\pi N_{i,o} kT_{i,o} R_b^3 \right) . \tag{11.32}$$

In order to avoid the explicit calculation of the reaction energy released during the fusion burn time, τ_b, as suggested by Eq. (7.15), let us take here the proportionality

$$E^*_{fu} \propto N^2_{i,o} V_b \tau_b \tag{11.33}$$

and combine this with Eqs. (11.29) and (11.30) to obtain

$$M_p \left(4\pi N_{i,o} kT_{i,o} R^3_b \right) \propto \eta_c N^2_{i,o} \cdot \tfrac{4}{3} \pi R^3_b \cdot \tau_b . \tag{11.34}$$

Further, recalling the expression for the disassembly speed, Eq.(11.11), and relating

$$\tau_b = \tau_{dis} \tag{11.35}$$

we re-arrange Eq. (11.34) to isolate R_b as

$$R_b \propto \left(\frac{M_p}{\eta_c} \right) \cdot \frac{kT_{i,o} \, v_{dis}}{N_{i,o}} . \tag{11.36}$$

The beam energy requirement now follows by substituting this expression for R_b in Eq. (11.32) to yield

$$E^*_{be} = A_{be} \frac{M^3_p}{N^2_{i,o} \eta^4_c} \tag{11.37}$$

where A_{be} is a temperature dependent factor. The exponents here are significant since they indicate a beam energy sensitive to $N_{i,o}$, η_c and M_p. For the case of a target at normal liquid density and with $M_p \approx 100$, and even for $\eta_c \approx 1$, this energy is found to be of the order of 10^{12} J, which is inaccessibly large. (Note that relating this to the approximately 10^{-9} s during which this energy needs to be delivered would represent an input power greatly exceeding the total steady power capacity of all electric power plants in North America.) If, however, target compression increases the density by 10^3, then the energy requirement is reduced by 10^6. This illustrates the reason for the very strong interest in high compression of the target.

11.5 Beams and Targets

In the discussion up to this point, little reference to the specifics of beams or target design has been made. As indicated, however, the requirements of pellet compression and beam-target coupling impose some very stringent demands on beam energy and the details of pellet composition.

High powered glass lasers have been used most frequently in inertial confinement fusion (ICF) research. The available beam intensity, focusing capability, state of technological development, and general availability are responsible for this popularity. Table 11.1 contains some properties of current lasers and, for comparison, also lists estimated requirements for actual fusion reactors.

Concerns about the eventual applicability of lasers to inertial confinement fusion has prompted considerable research on the potential role of ion

Parameter	Laser Type		
	Nd	KrF	Required
Wavelength (mm)	1.06	0.25	~ 0.3
Pulse rate (Hz)	0.001	5	~ 5
Beam energy (MJ)	0.03	0.1	~ 1
Representative peak power (TW)	30	100	~ 1000

Table 11.1: Status of Current Laser Technology and Requirements

accelerators for such purposes. The prospects of enhanced localized beam energy deposition and control over beam energy while avoiding the previously mentioned undesired preheating by "hot" electrons produced in the laser light absorption process makes this alternative very promising. However, accelerators do introduce another set of problems among which are beam focussing for high-current accelerators as well as the need for large high-vacuum ion transport facilities. Table 11.2 lists some ion accelerator characteristics.

Parameter	Accelerator Type			
	Electron	Light Ions	Heavy Ions	Required
Beam particle	e^-	p, α, C^{4+}	Xe, ..., U	–
Particle energy (MeV)	~ 10	~ 50	~ 30 000	> 10
Beam energy (MJ)	1	1	5	~ 5
Peak power (TW)	20	20	200	~ 1000

Table 11.2: Status of Current Accelerator Technology and Requirements

The composition of a pellet constitutes some interesting analysis and design problems. The existing types can be loosely grouped into three categories: (1) glass microballoons, (2) multiple shell pellets and (3) high-gain ion beam pellets.

Glass microballoons consist of thin walled glass shells containing a D_2-T_2 gas under high pressure, Fig. 11.3. The incident beam energy is deposited in the glass shell causing it to explode with part of its mass pushing inward and the remaining mass outward. Though microballoons are widely used in experiment, more efficient designs will eventually be needed for power plants.

Multiple-shell pellets contain an inner deuterium-tritium solid fuel core surrounded by a high-Z inner pusher-tamper. Next is a thicker layer of low density gas surrounded by a pusher layer. Finally, a low-Z ablator material forms the outer layer, Fig. 11.3. This complex layer structure is designed for very specific functions. For example, the outer layer is to ablate quickly and completely when struck by the incident beam; the inner high-Z pusher-tamper is to shield the inner core region against preheating by hot electrons and X-rays.

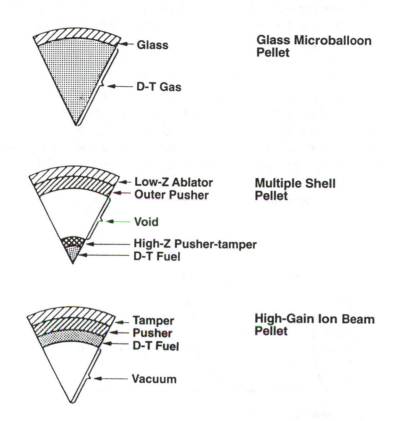

Fig. 11.3: Cross section of selected pellets for inertial confinement fusion.

More recently, heavy ion beam-pellets have been developed which depart in significant ways from the microballoon and layer shell design. In these designs, a vacuum sphere is surrounded by a D_2-T_2-DT fuel shell which is then surrounded by tamper-pusher materials, Fig.11.3. These pellets are designed specifically for ion beam inertial confinement fusion with the thickness of the tamper-pusher materials carefully matched to the type and energy of the incident beams.

Major objectives of these designs are to optimize energy transfer, minimize hot electron production, and reduce requirements for symmetric beam energy deposition.

11.6 Indirect Drive

The development of beam-target configurations, specifically the effort to achieve increased symmetric energy deposition over the surface of the pellet, has led to what is known as indirect drive. The distinction between this approach and that of direct drive–imparting pulses of energy from lasers or ion beams directly onto the pellet as previously discussed–is as follows.

For a system utilizing indirect drive, the target consists of both a fuel pellet– similar to those discussed in Sec.11.5–and a small cylindrical cavity, inside which the pellet is located. This cylindrical vessel, known as a "hohlraum", is a few cm long, is made of a high-Z material such as gold or other metal, and has "windows" transparent to the driver on each end, Fig.11.4. Then, instead of requiring all the driver beams to impinge symmetrically on the pellet, as is necessary for direct drive pellet compression, the beams enter both ends of the hohlraum obliquely and ablate the inner surface of the cavity. The high-Z material of the hohlraum emits soft X-rays when so irradiated, and by focusing the driver beams to the appropriate points inside the cavity, a highly symmetric irradiation of the fuel pellet results–followed by the previously discussed stages of pellet compression and heating depicted in Fig. 11.1.

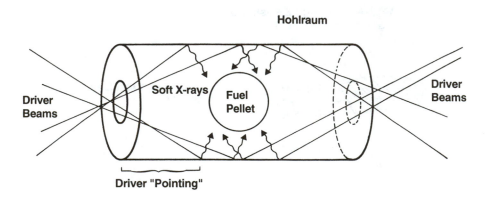

Fig. 11.4: Simplified depiction of an ICF indirect drive target system, including the hohlraum, fuel pellet, and driver beams with the resulting X-rays.

As with most methods of ICF pellet compression to date, indirect drive has only been examined using laser drivers, however, the same approach is believed to be applicable to ion beams. The crucial aspect of indirect drive is establishing the optimal "pointing" of the laser beams–the positioning and focusing of the beams on the cavity's inner surface which results in symmetric irradiation of the fuel pellet by the emitted soft X-rays. Development has shown this not only to be possible, but symmetric energy deposition on the pellet surface is achieved with fewer complications than when all the driver beams must symmetrically impinge directly on the pellet.

The other major advantages of indirect drive as compared to direct drive are the better ablation and subsequent compression achieved with X-rays as opposed to the visible light of lasers, and reduced instabilities during pellet compression. These characteristics and the demonstration of energy deposition levels over an entire pellet surface with $\leq 1\%$ deviation from uniformity are very appealing. However, the reduced energy coupling from the beam to the pellet, η_c in Eq.(11.17), and the increased complexities of hohlraum manufacture–in addition to the pellets alone–when scaled up to a power plant-type system are disadvantages not be overlooked. Despite these drawbacks, the inclusion of the hohlraum concept and the use of indirect drive-pellet compression does appear necessary in the continuing development of inertially confined fusion systems.

The designer of an ICF power plant faces a number of key decisions including the choice of the driver and the choice of protection scheme for the reaction chamber wall. The selection of a driver will depend on advances in the technology associated with specific types of drivers and on the beam-target coupling efficiency that can be achieved with specific beams. Protection for the reaction chamber wall from radiation and pellet debris released in a microexplosion is a unique and challenging aspect of ICF reactor design. Various possible approaches have been proposed including a large radius chamber with a "dry" wall and various "wet" wall concepts such as a falling liquid metal veil, liquid metal jets, liquid metal droplet sprays or a thin surface layer of liquid metal. The latter concepts all allow a smaller, more compact chamber but face various problems such as the difficulty in quickly pumping out vaporized material between pulses. Other unique design issues relate to pellet manufacture, pellet handling and positioning in the chamber, protection of mirrors, focusing magnets for ion beams, and other beam transport elements.

11.7   General Layout and Operation

The unique design features of an ICF power system relate to the reaction chamber and ancillary components. Underlying these features–with critical consequences on design criteria–are the nuclear-atomic energetics processes leading to fusion with the evident requirement of economic and self-sustaining performance of the system.

Fuel pellet manufacture involves spherical coating technology at the micro-scale of composition and geometry. Entry of the pellets into the burn chamber will occur by gravity combined with pneumatic injection demanding, however, extreme trajectory precision. Then, an inordinate quantity of energy has to be deposited into this small pellet by laser or ion beam impingement within the short time of about 10^{-9} s. The expected multiplied quantity of energy over the burn time $\tau_b = 10^{-8}$ s, now residing in the high kinetic energy of the fusion reaction products as well as in various electromagnetic flows and an assortment of debris, will spread out striking a liquid or metallic first wall surface. Both surface and internal radiation damage will occur as well as energy deposition–which needs to be recovered at the average rate that it is deposited. Simultaneously, tritium breeding by neutron capture will occur providing thereby eventual replacement of the scarce tritium fuel. Further aspects of the processes and reaction involved in the blanket surrounding the fusion chamber are addressed in Ch. 13.

Following each pulse, rapid purging of the reaction chamber needs to be undertaken in preparation for the next pulse. This operational cycling is expected to be at a frequency of about 1 Hz or greater, the associated fuel injection rate correspondingly being F_{+i}^* as previously introduced.

Extremes of power transport, energy conversion, material flow, radiation damage, and highly co-ordinated electro-mechanical functions will evidently characterize the eventual operation of an ICF power system. Considerable research, design, and testing will still need to be undertaken to arrive at the continuingly elusive goal of such a working power station.

Problems

11.1   Evaluate τ_b, Eq.(11.10) as a function of f_b for kT = 20 keV; take $\rho_b = 500$ ρ_ℓ.

11.2   Discuss the averaging process for $\overline{<\sigma v>}_{dt}$ and explain the approximation for it made to evaluate Eq.(11.15).

11.3   Derive a relation between ρR and the compression ratio, $\rho_b < \rho_\ell$, of a simple spherical target. Is a spherical target (e.g. microballoon) target advantageous compared to a disk or planar target?

11.4   Calculate the laser energy required to heat a spherical 50:50% D-T pellet, which attains $\rho_b R_b = 3$ g·cm^{-2}, to an average kinetic temperature of kT = 20 keV as a function of the pellet density ρ assuming that only 5% of the laser light is absorbed in the pellet fusion plasma. Evaluate this expression for the cases of
    (a) solid density (frozen state) $\rho_s \approx 0.22$ g·cm^{-3}
    (b) $\rho_b \approx 10^4 \rho_s$
and ascertain the corresponding pellet radius in each case as well as the respective confinement times, τ_{ic}, the laser power requirements and the fusion energy release for a 10% burnup fraction.

11.5   Formulate the exact calculation of E_{fu}^* in Eq.(11.33). Can the suggested proportionality be validated?

11.6   Undertake an analysis of the difference between the physical processes involved in energy deposition by laser beams and ion beams.

12. Low Temperature Fusion

Magnetic and inertial confinement approaches to controlled nuclear fusion have been shown to involve heating the fuel to high temperature and then confining it long enough for a sufficient quantity of fusion energy to be generated; high temperatures are required in order to counter the effect of Coulomb repulsion among the fuel ions. In contrast, the attainment of fusion energy at low temperature is based on the notion that the effect of Coulomb repulsion can be significantly reduced by a selective and temporary state of pseudo-charge neutrality among the fusile reactants.

12.1   Low and High Temperature Reactions

The well known neutron-induced fission of a uranium-235 nucleus, commonly written as

$$n + {}^{235}U \rightarrow \nu n + P_1 + P_2 \qquad (12.1)$$

is known to proceed at room temperature because one of the reactants is a neutral particle and a uranium-235 nucleus possesses a substantial fission cross-section for thermal neutrons. The absence of Coulomb forces of repulsion and the presence of nuclear forces of attraction at short distances of separation ($\leq R_o$) suggests a potential energy diagram as depicted in Fig.12.1a.

In contrast to the above, fusion of a deuterium ion with a tritium ion is represented by

$$d + t \rightarrow n + \alpha \qquad (12.2)$$

but requires a high reactant temperature to allow a sufficient number of ions to overcome the Coulomb barrier, or to penetrate it by tunnelling. This will lead to substantial reaction rates and the consequent energy yield. The corresponding ion-ion fusion potential energy diagram is shown in Fig.12.1b.

These two potential energy diagrams, Fig.12.1a and 12.1b, represent two conceivable extremes. A case between these extremes can be conceived of by the following conceptualization for deuteron-triton fusion. Consider a deuterium atom and a catalytic tritium nearby with both particles at low kinetic energy of relative motion, that is in a medium of low temperature. The catalytic tritium is taken to consist of the usual nucleus–a proton and two neutrons–but rather than its normal electron in a Bohr orbit, it contains a catalyst x in an orbit very close to the nucleus. This particle x is expected to possess an electric charge so as to render the catalytic particle neutral. We add that this particle x may or may not be stable against radioactive decay and may or may not be in a stable tight orbit around the nucleus. As for any two approaching hydrogen atoms, here the deuterium atom and the catalytic tritium will tend to combine by hydrogen molecule formation, which accounts for the range of attraction outside R_o in Fig. 12.1c. To the neighbouring deuteron, the catalytic tritium will, during the lifetime of the catalytic state, appear like an oversize neutron; the two may thus form a compound ionic state where the deuteron and triton are close enough for nuclear forces of attraction to dominate and therefore render fusion at low temperature. In Fig.12.1c, a "Coulomb sliver" occurs at the distance of the catalyst's orbit and is expected to be thin enough to be penetrated on account of the nuclei's available energy in the molecularly bound state.

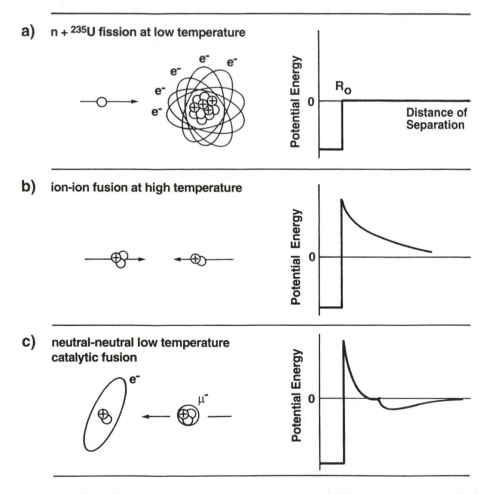

Fig. 12.1: Graphical depiction of low temperature fission, high temperature ion-ion fusion, and low temperature catalytic fusion.

The sequence of events for this low temperature catalytic fusion event consist of three distinct stages:

1. catalytic atom formation:
$$x + t \rightarrow xt \qquad (12.3a)$$

2. unstable intermediate formation:
$$xt + d \rightarrow xtd \qquad (12.3b)$$

3. decay into fusion reaction products:
$$xtd \rightarrow n + \alpha + x. \qquad (12.3c)$$

These three stages may also be written in sequential form
$$x + t + d \rightarrow xt + d \rightarrow xtd \rightarrow n + \alpha + x \qquad (12.3d)$$
and evidently possess some similarity to a fission process which, with a more detailed accounting of the process of reaction (12.1), may be written as

$$n + {}^{235}U \rightarrow {}^{236}U^* \begin{cases} {}^{236}U + \gamma \\ \nu n + P_1 + P_2. \end{cases} \qquad (12.4)$$

Indeed, we will show that branching reaction channels, each with their own probability, shown here in fission also apply to the catalytic reaction chain of Eq.(12.3d).

Low temperature fusion for which the catalyst x is a muon–recall our discussion of Sec. 7.7–has been experimentally demonstrated in liquid media at elevated pressures and in the temperature range 300 K to 900 K, formidable to muonic molecular formation. Evidently, if the process can be sustained as energetically and technologically favourable, then this novel approach might become a contender for a fusion device.

Some elementary aspects of muon physics can be described by the following. We begin with muon production. It is known that many high energy nuclear reactions yield the negative pi meson, π^-, as a reaction product,

$$\left(\text{High Energy Reaction}\right) \rightarrow \pi^- + \dots. \qquad (12.5a)$$

This pion possesses a mean life of $\sim 10^{-8}$ s and decays via

$$\pi^- \rightarrow \mu^- + \bar{\nu}_\mu \qquad (12.5b)$$

where $\bar{\nu}_\mu$ is the muon antineutrino. The negative muon μ^- decays with a mean life of 2.2×10^{-6} s according to

$$\mu^- \rightarrow e^- + \bar{\nu}_e + \nu_\mu \qquad (12.6)$$

where e^- is an electron, $\bar{\nu}_e$ is an electron antineutrino, and ν_μ is the muon neutrino. It is common to dispense with the adjective "negative" for the muon and simply represent this particle by μ rather than μ^-.

The initial kinetic energy of a produced muon depends upon the details of the initiating reaction but is typically about 200 MeV. In a dense liquid hydrogenous medium, this high energy subatomic particle slows down to about 2 keV in $\sim 10^{-8}$ s and in another $\sim 10^{-11}$ s cascades down into a K-orbit around a deuteron or triton to form a muonic atom, μd or μt.

The details of the subsequent muon-nucleus, muon-atom and muon-molecule interactions are complex; for example, resonance phenomena involving muonic atoms and associated molecule formations have been identified suggesting the appearance of a variety of nuclear and atomic states. However, for present purposes, and in order to illustrate low temperature fusion, we incorporate these various processes in a collective dynamic characterization using macroscopic reaction parameters.

12.2   Compound Decay Reaction

In order to emphasize some essential concepts involving multi-stage low temperature reactions, we first consider the following general reaction process involving arbitrary a-type and b-type particles:

$$a + b \rightarrow (ab) \rightarrow c + d. \tag{12.7}$$

Here, the formation of the intermediate (ab) comes about as a result of a binary collision process and this intermediate is subsequently transformed by a decay process. Thus, though the same straight arrow symbol is used for the formation and decay processes in the reactions of Eq.(12.7), the two constitute distinctly different physical phenomena.

We introduce a reaction parameter κ_{ab} for the formation of the intermediate species (ab) in reaction (12.7) so that the rate density of formation of the new species (ab) of density N_{ab} is given by

$$R_{+ab} = \kappa_{ab} N_a N_b. \tag{12.8}$$

Thus, if this formalism is applied to the d-t fusion process, we can equate $\kappa_{dt} = \langle \sigma v \rangle_{dt}$.

The decay of (ab) in reaction (12.7) is equivalent to the process of radioactive decay of an (ab) particle of density N_{ab}. Thus, the decay rate density can be characterized by a mean lifetime $1/\lambda_{ab}$ for the intermediate, giving therefore

$$R_{-ab} = \lambda_{ab} N_{ab}. \tag{12.9}$$

By inspection of the reactions in Eq.(12.7), the rate equations for the various particles a, b, ab, c and d in terms of their densities N_a, N_b, N_{ab}, N_c and N_d are given by the following:

$$\frac{dN_a}{dt} = -\kappa_{ab} N_a N_b \tag{12.10a}$$

$$\frac{dN_b}{dt} = -\kappa_{ab} N_a N_b \tag{12.10b}$$

$$\frac{dN_{ab}}{dt} = \kappa_{ab} N_a N_b - \lambda_{ab} N_{ab} \tag{12.10c}$$

$$\frac{dN_c}{dt} = \lambda_{ab} N_{ab} \tag{12.10d}$$

$$\frac{dN_d}{dt} = \lambda_{ab} \, N_{ab} \, . \tag{12.10e}$$

With the imposition of initial conditions for each of the reacting species, the dynamical description of reaction (12.7) is thus fully specified. Note, however, that these equations are nonlinear and also display some redundancy.

The above characterization will now be used to provide a reduced dynamic description of the muon-catalyzed d-t fusion process. As an initial simplified case, if muon decay and its parasitic loss by alpha capture are ignored, then a typical muon-catalyzed chain would be written, with x in the reactions of Eqs. (12.3) replaced by μ, as

$$\mu + t \rightarrow \mu t \tag{12.11a}$$

$$\mu t + d \rightarrow \mu dt \tag{12.11b}$$

$$\mu dt \rightarrow n + \alpha + \mu \, . \tag{12.11c}$$

These three reactions proceed at the rate densities

$$R_1 = \kappa_{\mu t} \, N_\mu \, N_t \tag{12.12a}$$

$$R_2 = \kappa_{\mu t d} \, N_{\mu t} \, N_d \tag{12.12b}$$

and

$$R_3 = \lambda_{\mu dt} \, N_{\mu dt} \tag{12.12c}$$

where the notation $\kappa_{(\,)}$, $\lambda_{(\,)}$ and $N_{(\,)}$ is used in the sense previously defined. This reaction can be graphically suggested in the following sequential form:

$$
\begin{array}{ccccccc}
 & R_1 & & R_2 & & R_3 & \\
\mu + t & \rightarrow & \mu t & \rightarrow & \mu dt & \rightarrow & n + \alpha + \mu \, . \\
\end{array}
\tag{12.13}
$$

with rates $\kappa_{\mu t}$, $\kappa_{\mu t d}$, $\lambda_{\mu dt}$ and recycling of the muon.

Here the chain sustainment by recycling of the muon is evident. This representation, however, does not include the decay of the muon which is known to occur regardless of whether the muon is free or bound to a nucleus. Additionally, it does not explicitly describe the possibility of a muon, freed upon a fusion reaction, sticking to an alpha particle produced therein, which occurs with a probability ω. This latter process is particularly crucial to the catalytic cycle since a muon stuck to an alpha cannot be of use for reaction propagation. These two additional processes, muon decay and muon sticking, lead to the following more inclusive, though still incomplete, reaction network:

$$
\begin{array}{c}
\lambda_\mu \quad\quad \lambda_\mu \quad\quad \lambda_\mu \quad\quad\quad\quad \lambda_\mu \\
\mu + t \;\rightarrow\; \mu t \;\rightarrow\; \mu dt \xrightarrow{\;\;\omega\;\;} \mu\alpha + n \\
\kappa_{\mu t} \quad\quad \kappa_{\mu t d} \quad\quad \lambda_{\mu dt} \quad 1-\omega \searrow \alpha + \mu + n
\end{array}
\tag{12.14}
$$

where λ_μ is the muon decay constant. The reaction shown here represents the dominant processes involved in muon-catalyzed d-t fusion. However, a number of additional reactions are possible as we display in Fig.12.2.

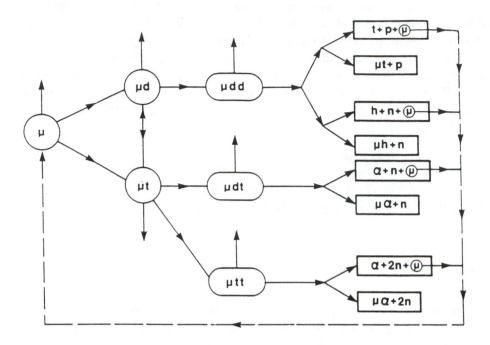

Fig. 12.2: Comprehensive network of possible muon paths in a d-t liquid medium. Decay of the muonic reaction products is not shown.

12.3   Energy Viability

An assessment of the energy liberated in a nuclear reaction–or in a sequence of reactions–relative to the energy cost of causing that reaction, is fundamental for evaluating the attractiveness of any nuclear energy system. Recall that such a criterion was employed in the energy balance assessment for magnetic and inertial confinement fusion. We now seek to formulate a similar energy balance criterion for a muon-catalyzed d-t system, i.e. a µdt system.

One unique feature of muon-catalyzed fusion is that muons have to be produced by appropriate accelerators and then directed into a liquid deuterium-tritium mixture. The accelerator beam energy requirements are such that the average energy cost of a muon is of the order of 3000 MeV. Then, since each d-t fusion reaction catalyzed by a muon releases the usual 17.6 MeV of fusion energy, energy breakeven will require each muon to catalyze, on average, several hundred d-t fusion reactions during its short lifetime of 2.2×10^{-6} s. Consequently, the rates at which the various processes of Fig. 12.2 occur are most important.

To establish a tractable formulation for this energy balance problem, we consider a unit volume of liquid deuterium and tritium into which muons are injected at a rate density F_μ, Fig.12.3. These muons then initiate and sustain a complex reaction network as suggested in Fig.12.2. The fusion energy so generated heats the fluid-fuel mixture; this heat is transported with the moving fluid to heat exchangers for subsequent conversion. For the muon-sustained reactions, it is known that an acceptable loss in accuracy results if the reduced reaction network of Fig.12.4 is used instead of that of Fig.12.2.

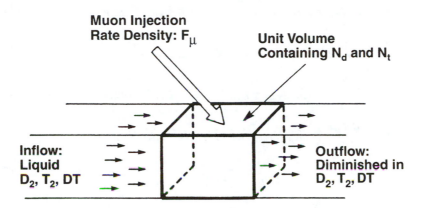

Fig. 12.3: A beam of muons entering a unit volume of flowing liquid deuterium-tritium.

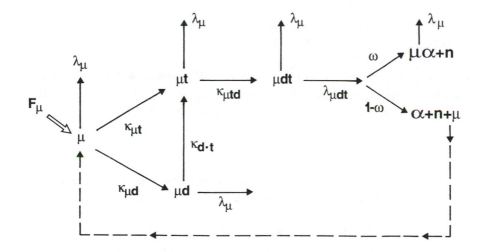

Fig. 12.4: Reduced μ-d-t reaction network. The symbol ω represents the probability of the muon sticking to the alpha particle and the other symbols are used or defined in the text.

The densities of the various nuclear/atomic/molecular species in a considered reaction unit volume can then be formulated by inspection–based on our discussion of Sec.12.2–and vary with time according to the following system of rate equations:

$$\frac{dN_\mu}{dt} = -\lambda_\mu N_\mu - \kappa_{\mu t} N_\mu N_t - \kappa_{\mu d} N_\mu N_d + \lambda_{\mu dt} N_{\mu dt}(1-\omega) + F_\mu \tag{12.15a}$$

$$\frac{dN_t}{dt} = -\kappa_{\mu t} N_\mu N_t - \kappa_{d\cdot t} N_{\mu d} N_t + \lambda_\mu N_{\mu t} + F_{+t} - F_{-t} \tag{12.15b}$$

$$\frac{dN_d}{dt} = -\kappa_{\mu d} N_\mu N_d + \kappa_{d\cdot t} N_{\mu d} N_t + \lambda_\mu N_{\mu d} + F_{+d} - F_{-d} - \kappa_{\mu dt} N_{\mu t} N_d \tag{12.15c}$$

$$\frac{dN_{\mu t}}{dt} = -\lambda_\mu N_{\mu t} + \kappa_{\mu t} N_\mu N_t + \kappa_{d\cdot t} N_{\mu d} N_t - \kappa_{\mu dt} N_{\mu t} N_d \tag{12.15d}$$

$$\frac{dN_{\mu d}}{dt} = -\lambda_\mu N_{\mu d} + \kappa_{\mu d} N_\mu N_d - \kappa_{d\cdot t} N_{\mu d} N_t \tag{12.15e}$$

$$\frac{dN_{\mu dt}}{dt} = -\lambda_\mu N_{\mu dt} + \kappa_{\mu dt} N_{\mu t} N_d - \lambda_{\mu dt} N_{\mu dt} \tag{12.15f}$$

$$\frac{dN_n}{dt} = \lambda_{\mu dt} N_{\mu dt} \tag{12.15g}$$

$$\frac{dN_\alpha}{dt} = \lambda_{\mu dt} N_{\mu dt}(1-\omega) + \lambda_\mu N_{\mu\alpha} \tag{12.15h}$$

$$\frac{dN_{\mu\alpha}}{dt} = \lambda_{\mu dt} N_{\mu dt}(\omega) - \lambda_\mu N_{\mu\alpha} . \tag{12.15i}$$

Here, in Eq. (12.15b), the factor $\kappa_{d\cdot t}$ is to account for the net reaction transfer from μd to μt, also shown in Fig. 12.4. These equations represent a so-called point-kinetics representation in the sense that any spatial effects can be considered to be minimal.

By analogy to the energy viability analysis of magnetic and inertial systems, we consider a reactor chamber of volume V_r and an operating period τ during which muon injection occurs. Then, the total energy supplied is

$$E_{in}^* = V_r \int_0^\tau F_\mu(t) E_{\mu c} dt = V_r E_{\mu c} \int_0^\tau F_\mu(t) dt \tag{12.16}$$

where $E_{\mu c}$ is the average energy cost of producing one muon. The total energy released as a consequence of muon injection can be written as

$$E_{out}^* = V_r \int_0^\tau F_\mu(t) E_{\mu s} dt + V_r \int_0^{\tau+\tau_\mu} \left(-\frac{dN_{\mu dt}}{dt}\right)_{fusion} Q_{dt} dt \tag{12.17}$$

where now $E_{\mu s}$ is the average energy deposited in the medium per muon as the high energy muons are slowing down. Note the addition of the muon mean lifetime τ_μ to the injection period; this is necessary because a muon will continue to catalyze nuclear reactions during its lifetime after injection has ceased.

Since the reaction energy Q_{dt} is released whenever the (μdt)-fusion compound decays, we write for the fusion rate density in Eq.(12.17)

$$\left(-\frac{dN_{\mu dt}}{dt}\right)_{fusion} = \lambda_{\mu dt} N_{\mu dt}(t) \tag{12.18}$$

so that Eq.(12.17) becomes

$$\frac{E^*_{out}}{V_r} = E_{\mu s} \int_0^\tau F_\mu(t)dt + Q_{dt} \int_0^{\tau+\tau_\mu} \lambda_{\mu dt} N_{\mu dt}(t)dt . \tag{12.19}$$

The energy multiplication ratio, M_E, is given by

$$M_E = \frac{E^*_{out}}{E^*_{in}} = \left(\frac{Q_{dt}}{E_{\mu c}}\right)\frac{\int_0^{\tau+\tau_\mu} \lambda_{\mu dt} N_{\mu dt}(t)dt}{\int_0^\tau F_\mu(t)dt} + \frac{E_{\mu s}}{E_{\mu c}} . \tag{12.20}$$

Here, the ratio $E_{\mu s}/E_{\mu c}$ accounts for the recoverable muon beam energy and is of the order of 1/10. Of greater interest therefore is the first term of Eq.(12.20) where Q_{dt} and $E_{\mu c}$ are constants and the remaining factor defines the important parameter χ_μ which represents the average number of d-t fusion events catalyzed by one muon:

$$\chi_\mu = \frac{\int_0^{\tau+\tau_\mu} \lambda_{\mu dt} N_{\mu dt}(t)dt}{\int_0^\tau F_\mu(t)dt} . \tag{12.21}$$

This parameter χ_μ, called the muon recycle efficiency, is evidently of utmost importance and will be considered in detail next.

12.4   Muon Catalysis Efficiency

A literal interpretation of the parameter χ_μ is that it represents the average number of d-t fusions a muon can catalyze during its lifetime. Equation (12.21) could be evaluated if the muon injection rate density $F_\mu(t)$ is known as a function of time together with the time dependent concentration $N_{\mu dt}(t)$; the former is specified by accelerator operation but the latter can only be determined from a solution of the system of dynamical particle balances given by Eqs.(12.15) in combination with the energy dynamics in the reaction domain. To avoid having mathematical complexity obscure the physical features of the problem, we will assume steady-state operation. That is, we assume a constant fuel temperature allowing therefore a constant reaction rate parameter and also take a constant injection rate, $F_\mu(t) = F_\mu^o$ during the operating time τ of interest. Hence, after an initial transient during start-up, the density of the $N_{\mu dt}(t)$ molecular ions reaches a constant value, $N_{\mu dt}^o$, characteristic of steady-state. Then the muon recycle efficiency χ_μ, Eq.(12.21), becomes

$$\chi_\mu = \frac{\lambda_{\mu dt} N_{\mu dt}^o (\tau + \tau_\mu)}{F_\mu^o(\tau)} \approx \lambda_{\mu dt} \left(\frac{N_{\mu dt}^o}{F_\mu^o} \right). \tag{12.22}$$

Note that the reactor operating time τ will invariably be much in excess of the muon mean lifetime so that $\tau + \tau_\mu \approx \tau$.

By maintaining constant fuel and muon densities by appropriate feed rates, a constant $N_{\mu dt}$ implies that all the other intermediate particle densities will also be constant in time; that is, we will have in Eqs.(12.15)

$$\frac{dN_\mu}{dt} = \frac{dN_t}{dt} = \frac{dN_d}{dt} = \frac{dN_{\mu t}}{dt} = \frac{dN_{\mu d}}{dt} = \frac{dN_{\mu dt}}{dt} = 0. \tag{12.23}$$

Under these conditions, the system of linear algebraic equations for N_μ, $N_{\mu d}$, $N_{\mu t}$ and $N_{\mu dt}$, Eq.(12.15a) and Eqs.(12.15d) to (12.15f), can be solved to yield an explicit ratio $N_{\mu dt}^o / F_\mu^o$ as required for Eq.(12.22). This gives

$$\frac{N_{\mu dt}^o}{F_\mu^o} = \frac{1}{\lambda_{\mu dt}(\omega + B_\mu)} \tag{12.24a}$$

where

$$B_\mu = \frac{\left(\lambda_\mu + \kappa_{\mu t} N_t^o + \kappa_{\mu d} N_d^o\right)\left(\lambda_\mu + \kappa_{\mu dt} N_d^o\right)\left(\lambda_\mu + \lambda_{\mu dt}\right)}{\lambda_{\mu dt} \kappa_{\mu dt} N_d^o \left(\kappa_{\mu t} N_t^o + \dfrac{\kappa_{\mu d} \kappa_{d \cdot t} N_d^o N_t^o}{\lambda_\mu + \kappa_{d \cdot t} N_t^o} \right)} - 1 \tag{12.24b}$$

and may be interpreted as a muon "residence unavailability" penalty. In compact form the muon recycling efficiency therefore reduces to

$$\chi_\mu = \left(\frac{1}{\omega + B_\mu} \right). \tag{12.25}$$

For the case of muon catalysis at liquid hydrogen conditions, Table 12.1, the muon recycle efficiency is calculated to be

$$\chi_\mu \approx 34. \tag{12.26}$$

That is, on average, one muon catalyzes some 34 d-t fusions during its mean lifetime of 2.2×10^{-6} s.

The energy multiplication assessment follows similarly from a substitution of Eq.(12.21) into Eq.(12.20); we take $Q_{dt} = 17.6$ MeV and use an estimate of $E_\mu \approx$ 3000 MeV to find

Process	Parameter	Value
Muon decay	λ_μ	0.45×10^6 s^{-1}
μ-d-t fusion	$\lambda_{\mu dt}$	1.1×10^{12} s^{-1}
Muonic atom formation	$\kappa_{\mu d}$	1.2×10^{-12} cm^3 s^{-1}
Muonic atom formation	$\kappa_{\mu t}$	1.2×10^{-12} cm^3 s^{-1}
Muonic molecule formation	$\kappa_{\mu dt}$	0.25×10^{-14} cm^3 s^{-1}
Isotope exchange process	$\kappa_{d\cdot t}$	0.5×10^{-14} cm^3 s^{-1}
Muon sticking probability	ω	0.007

Table 12.1: Parameters for the μ-d-t process of Fig.12.4. The numerical values used in this table are for illustrative purposes and may not correspond to the latest measurements. The fuel medium is taken to be a liquid deuterium-tritium mixture at 300 K ($N_d = N_t = 10^{22}$ cm^{-3}).

$$M_E = \left(\frac{Q_{dt}}{E_{\mu c}}\right)\chi_\mu + \frac{E_{\mu s}}{E_{\mu c}}$$

$$\approx \left(\frac{17.6}{3000}\right)(34) + 0.1 \approx 0.3 .$$

(12.27)

This provides a useful estimate of the energy viability of a muon catalyzed fusion system which, as is evident, is too low by perhaps a factor of thirty.

Some additional considerations may, however, be introduced which suggest possible increases in M_E. Clearly Q_{dt} is a constant and $E_{\mu c}$ could not be further reduced except for possible alternative–and highly speculative–methods of muon production. Considerable research has in recent years been undertaken to determine if χ_μ, the average number of d-t fusions catalyzed by one muon, could be increased. Experiments have revealed that at specific temperatures and at significantly elevated pressures, values of $\chi_\mu \approx 160$ are possible–corresponding to an increase in the energy multiplication of Eq.(12.27) by a factor of about 3.5. However, while a definite dependence on medium temperature and composition has been established, it is not yet evident that an according optimization can be sufficient for energy viability.

There exists, however, another approach which can be summarized by the following. Supposing the fusion domain is surrounded by a neutron multiplication and breeding blanket domain, the function of which is to multiply the neutron by (n,xn) reactions and to breed both tritium and fissile fuel for companion fission reactors. (Fusion-fission hybrids and similar integrated systems will be discussed in greater detail in Chs.14 and 15.) The fission energy eventually thus generated might be considered a benefit. For example, if a fission energy credit of $Q_{fi,credit} \approx 150$ MeV per d-t fusion reaction were possible then Eq.(12.27) would give–neglecting the small beam energy recovery–a quantity of

$$M_E = \left(\frac{Q_{dt} + Q_{fi,credit}}{E_\mu} \right) (\chi_\mu)$$

$$\approx \left(\frac{17.6 + 150}{3000} \right)(160) \approx 9$$

(12.28)

to yield a significant positive energy balance. Such a hybrid system, based on muon-catalyzed fusion, would have to be evaluated on the same basis as other fusion based hybrid concepts and will be further discussed. A potential advantage of this system could be the relative simplicity of the reaction chamber; however, accelerators which produce sufficient numbers of collimated muons at an acceptable average energy cost per muon represent a significant design uncertainty.

12.5   Muon Catalyzed Reactor Concept

The design of a muon-catalyzed reactor system is dominated by the need for an on-line accelerator to produce pi-mesons which are collected and then decay into the desired muons in a domain of interest. In this respect, the muon catalyzed fusion power flow pattern resembles that of an inertial confinement fusion system, Fig.11.2, particularly with respect to the need for a significant recirculating power flow to the accelerator.

   While a generally accepted design has yet to emerge–and largely awaits the development of suitable accelerators yielding a sufficiently intense beam of muons–we can conceive of a generalized schematic such as depicted in Fig.12.5. Here, light ions (p, d or t) of energy in excess of ~1 GeV will strike a low atomic mass number target with the transmitted ions either recirculated by magnetic forces or allowed to impact upon one or more successive targets of increasing atomic mass. The pi-meson will be emitted with a highly anisotropic directional distribution to be collected for decay into muons. The muons thus produced would be collected and focused onto a cylindrical fusion core consisting of liquid deuterium and tritium under very high pressure and at a temperature ~10^3 K. This cylindrical fusion core will be surrounded by a blanket which serves, variously and as required, the functions of (i) tritium breeding, (ii) fissile fuel breeding, and (iii) energy removal.

   The technology required to develop this conceptual system has many similarities to that involved in other fusion concepts; however, the pion collector, the focusing of muons into the fusion core, and the efficient recovery of residual energy in the accelerator target are problems yet to be resolved.

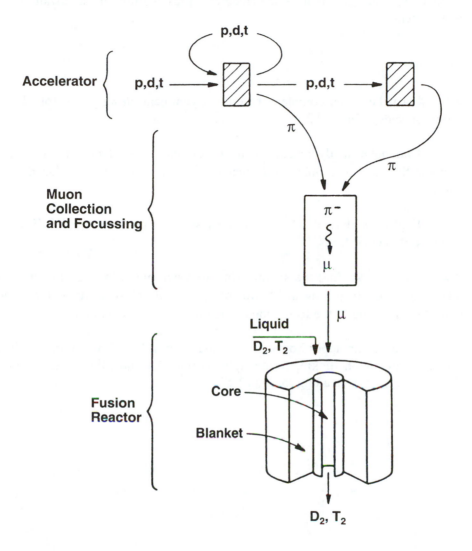

Fig. 12.5: Schematic showing the main components of a muon catalyzed d-t fusion reactor system.

12.6   Cold Fusion: A Comment

In this chapter we have focused on the concept of low temperature fusion based on the mechanism of muon catalysis. Recent years have witnessed considerable discussion of "cold" fusion in electrolytic cells and other experimental devices. No clear and consistent characterizations of the associated processes have emerged to date so that a pedagogical presentation must await further developments.

Problems

12.1   Formulate the complete reaction dynamical description for the 14-particle species of Fig. 12.2.

12.2   Undertake a dynamical analysis of the case depicted in Fig.12.4 assuming that $\lambda_\mu \approx 0$. Compare this characterization to a neutron induced fission chain.

12.3   Explore analytical solutions of the system of equations, Eq.(12.10), given suitable initial conditions.

12.4   For $\chi_\mu = 34$, find the average time between muon induced fusion events and compare this to the mean lifetime of the μdt atom. What does this suggest about the time scale of the formation of the μdt atom relative to its decay?

12.5   A particular upper bound for the muon recycle efficiency occurs in the absence of any muon sticking to the alpha particle. Examine the consequences of this limit for the case leading to Eq.(12.27).

PART IV COMPONENTS, INTEGRATION, EXTENSIONS

13. Fusion Reactor Blanket

The structure immediately surrounding the fusion reaction chamber needs to serve several functions, among which are the following: (i) to sustain a sufficiently clean plasma domain, (ii) to recover energy from the emitted radiation and reaction products, (iii) to shield the surrounding structures and personnel, and (iv) to breed tritium required in the d-t reactor core. This fusion reactor blanket thus serves a most essential role and deserves close examination. While the fusion chamber is generally maintained at plasma conditions such that– at best–an energy self-sufficient reaction chain is established, note that it is indeed the adherent blanket where the neutron and radiation energy released from the plasma is deposited, and which will finally provide the energy transformable for external utilization, i.e. for driving electric generators in a power plant.

13.1   Blanket Concept

Our previous discussion of magnetic confinement fusion (MCF), inertial confinement fusion (ICF) and muon catalyzed fusion makes it clear that each represents a different set of physical conditions requiring that the blanket design must be adapted to best suit each case.

The MCF plasma domain can be characterized by a low density gas ($\sim 10^{21}$ particles$\cdot$m^{-3}) at a high temperature ($\sim 10^8$ K). These properties lead to some very specific design and operation requirements. Evidently, all fueling and diagnostic wall penetrations into the core must sustain a laboratory vacuum. Then, the need for a high plasma temperature requires that the plasma must be separated from the wall surface. Concurrently, the intense neutron flux striking the wall demands that its composition be highly resistant to neutron damage. Finally, provisions have to be made for the removal of the neutron and radiation energy deposited in the blanket. Figure 13.1 suggests some of the essential components and features of an MCF blanket system.

Different conditions apply to an ICF device. In this case, reactor operation involves target pellets on a ballistic trajectory which are struck by extremely intense pulses of ions or electro-magnetic radiation (lasers) when the pellet reaches a specific location. As a consequence of the resulting ablation, compression, fusion, and disintegration of the target, an intense pressure and radiation shock wave is generated which must be absorbed in the first wall. A favoured approach is to protect the wall by a liquid metal so that the debris and radiation from the explosion are then absorbed in this layer, thus providing both shielding and a moving heat transfer fluid. We suggest this scheme in Fig.13.2.

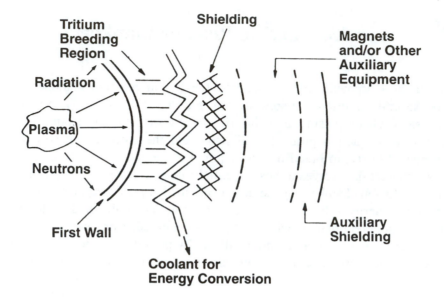

Fig. 13.1: Components of a blanket for a magnetic confinement fusion reactor.

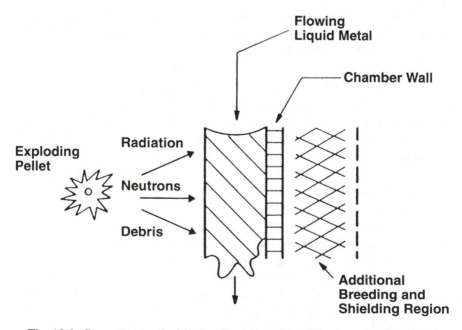

Fig. 13.2: Components of a blanket for an inertial confinement fusion reactor.

In contrast to the above MCF and ICF, a muon catalyzed deuterium-tritium (μDT) fusion reactor blanket domain may be depicted as suggested in Fig.13.3. Its dominant feature is a central channel containing a deuterium-tritium oxide mixture either in liquid or two-phase form. The reaction product alpha will be retained in the flow but the penetrating neutrons will enter the surrounding blanket region. The first wall may thus possess many of the properties of cladding associated with existing fission reactors.

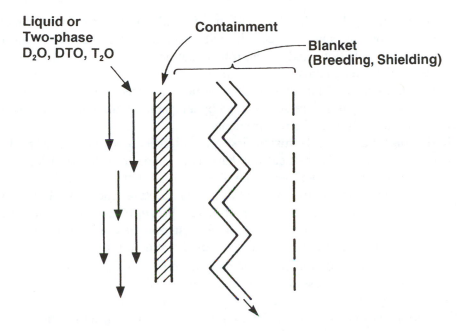

Fig. 13.3: Components of a blanket for a muon catalyzed deuterium-tritium fusion reactor.

A feature common to all deuteron-triton burning fusion systems is the need to breed tritium. Hence, the blanket domain must contain a concentration of lithium, either as a pure substance or in compounds, so that the neutrons emitted in d-t fusion are captured in the lithium to produce tritium. Such a nuclear reaction linkage may well be represented by the following:

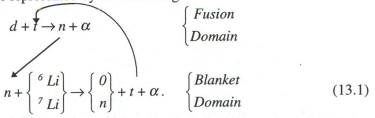

$$d + t \rightarrow n + \alpha \qquad \left\{ \begin{array}{l} Fusion \\ Domain \end{array} \right.$$

$$n + \left\{ \begin{array}{l} {}^{6}Li \\ {}^{7}Li \end{array} \right\} \rightarrow \left\{ \begin{array}{l} 0 \\ n \end{array} \right\} + t + \alpha . \qquad \left\{ \begin{array}{l} Blanket \\ Domain \end{array} \right. \qquad (13.1)$$

Should tritium be available from other sources, such as existing fission reactors, or if additional neutron multiplication is sustained by incorporating suitable (n,xn)-type materials in the blanket, then there may be an excess of neutrons over that required for tritium breeding. These surplus neutrons could then be used to breed fissile fuel by capture in fertile nuclei. Such a system concept will be discussed in Ch. 15; for now we consider further details of the MCF, ICF and µDT blankets.

13.2 First Wall Loading

As suggested in Fig.13.1, the first wall in a typical MCF reactor must directly face the fusion plasma. Hence, it will intercept the fusion neutrons, bremsstrahlung radiation and cyclotron radiation, as well as any plasma constituents which leak across the outer magnetic field lines. Therefore, the wall must maintain structural integrity against particle and electromagnetic radiation damage as well as against stresses induced by temperature gradients and pressure forces associated with vacuum requirements.

Radiation damage in a d-t fusion first wall is largely associated with the 14.1 MeV fusion neutrons. These neutrons have two important effects: i) knock-on collisions that displace nuclides from their normal lattice positions causing internal voids in the microstructure and ii) neutron capture reactions which result in ^{4}He production and hence a build-up of helium pressure in the material lattice; the latter causes a volumetric swelling of the material since helium is relatively immobile and does not rapidly diffuse out of the structure except at very high temperatures. These (n,α) reactions typically have thresholds requiring neutrons of energy in excess of several MeV; thus, the swelling phenomenon is most pronounced in d-t fusion devices due to the high flux of 14 MeV neutrons.

Radiation damage effects may be contained by setting a limit on the neutron wall loading. The average loading, here represented by Λ_n, is defined as

$$\Lambda_n = \frac{f_n \int_{V_c} P_{fu}(\mathbf{r}) d^3 r}{A_w} \tag{13.2}$$

where $P_{fu}(\mathbf{r})$ is the fusion reaction power per unit volume, V_c is the fusion core volume, A_w is the total wall area and f_n is the fraction of the fusion energy carried by the neutrons; the commonly used units for this wall loading are MW·m^{-2}. For ICF, however, note that–due to the high density of the compressed pellet– neutrons can be absorbed to some amount in the dense fusion plasma. For reasons of simplicity, we neglect this affect here.

The above definition can be expressed in a more explicit form by using the appropriate equation for the power profile $P_{fu}(\mathbf{r})$. Adopting the simplified geometry of an axisymmetric torus with circular cross section, and assuming only a radial dependence for the fusion power, gives for Eq.(13.2)

$$\Lambda_n = \frac{f_n}{A_w} \int_0^{2\pi R_o} \int_0^a \int_0^{2\pi} P_{fu}(\ell, r, \theta) r \, d\theta \, dr \, d\ell$$

$$= \frac{f_n}{A_w} \int_0^{2\pi R_o} \int_0^a P_{fu}(r) 2\pi r \, dr \, d\ell$$

$$= \frac{f_{n,dt} \gamma}{A_w} 4\pi^2 R \int_0^a <\sigma v>_{dt} N_d N_t Q_{dt} r \, dr \tag{13.3}$$

$$= \gamma f_{n,dt} Q_{dt} \left(\frac{2\pi^2 a^2 R_o}{A_w} \right) \frac{2}{a^2} \int_0^a [<\sigma v>_{dt} (T(r))][N_d(r)][N_t(r)] r \, dr$$

in the case of d-t fusion. Here γ is the conversion factor of MeV/s to MW, a is the minor radius of the torus and R_o is its major radius. Note that it is the fuel ion densities as well as their temperature as a function of radius which enter as the determining space-dependent functions.

Limits on the neutron wall loading Λ_n depend on the specific design and are most commonly set by radiation damage as it affects the component's lifetime. Recent designs generally specify a range of 1 - 5 MW·m^{-2} for this parameter.

Additional considerations now need to be added. For example, the power associated with Λ_n is not absorbed in the first wall itself since most of the neutrons are transmitted more deeply into the blanket with little attenuation. Thus, actual thermal wall loading must be evaluated separately based on (i) the incident bremsstrahlung and cyclotron radiation, (ii) direct neutron interactions, and (iii) the interactions associated with backscattered neutrons. Since the radiation is largely absorbed near the front surface of the wall, the surface temperature is strongly dependent on this power flow. Indeed, surface heat fluxes over 1 MW·m^{-2} may be difficult to transmit without exceeding surface temperature limits set by vaporization pressure considerations.

13.3 Plasma-Wall Interactions

The first wall surrounding a fusion plasma is bombarded by both electromagnetic radiation and escaping plasma particles. Numerous effects such as temperature changes, thermal stresses, and erosion must be considered in wall design. One important effect we will discuss here is that of material erosion by sputtering. In this process, the incident ions or neutrals from the plasma possess sufficient energy to cause "billiard ball"-type collisions in the wall material, possibly leading to the ejection of wall atoms into the plasma, as suggested in Fig.13.4.

To gain some insight into this important plasma-wall interaction, consider the process suggested in Fig.13.5. A plasma ion of mass m_1 and velocity v_1 enters the first-wall and collides elastically with a stationary atom of mass m_2. Using the notation of Fig.13.5, we write the momentum and energy balances for such elastic events as

$$m_1 v_1 = m_1 v_1' \cos\theta_1 + m_2 v_2' \cos\theta_2 \quad \text{(component } \| \text{ to } \mathbf{v}_1) \tag{13.4a}$$

$$m_1 v_1' \sin\theta_1 = m_2 v_2' \sin\theta_2 \quad \text{(component } \perp \text{ to } \mathbf{v}_1) \tag{13.4b}$$

and

$$\tfrac{1}{2} m_1 v_1^2 = \tfrac{1}{2} m_1 \left(v_1' \right)^2 + \tfrac{1}{2} m_2 \left(v_2' \right)^2. \tag{13.5}$$

A maximum of energy is transferred to the first-wall atom if a head-on collision occurs; for this case $\theta_1 = \pi$ and $\theta_2 = 0$ and the maximum energy transferred $(\Delta E)_{max}$ is given by

$$\left(\Delta E \right)_{max} = \tfrac{1}{2} m_1 v_1^2 - \tfrac{1}{2} m_1 \left(v_1' \right)^2 = \tfrac{1}{2} m_1 v_1^2 \left[1 - \left(\frac{v_1'}{v_1} \right)^2 \right]. \tag{13.6}$$

Algebraic manipulation of Eqs. (13.4a) and (13.5) for $\theta_1 = \pi$ and $\theta_2 = 0$ allows the elimination of the speed ratio (v_1' / v_1) to give

$$\left(\Delta E \right)_{max} = \tfrac{1}{2} m_1 v_1^2 \left[\frac{4 m_1 m_2}{\left(m_1 + m_2 \right)^2} \right] = \frac{4 m_1 m_2}{\left(m_1 + m_2 \right)^2} E_{inc}, \tag{13.7}$$

where E_{inc} is the kinetic energy of the incident ion.

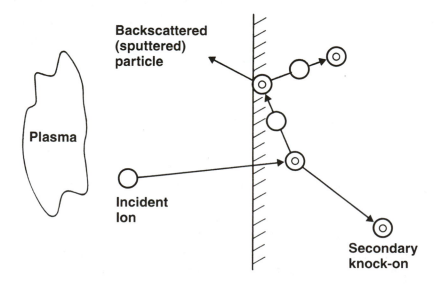

Fig. 13.4: Schematic diagram of typical ionic impact phenomena leading to sputtering.

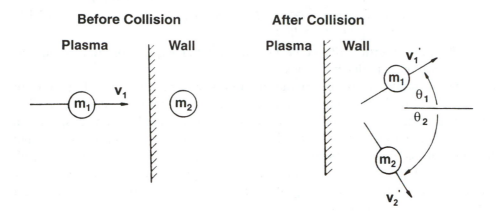

Fig. 13.5: Depiction of a plasma ion, m_1, elastically colliding with a first wall atom, m_2.

The other important parameter is the average distance a plasma ion moves into the first wall before it collides with an atom. Since all two-body collisions can be characterized by a microscopic cross-section, we use

$$\sigma_{col} = \pi R_{1,2}^2 \qquad (13.8)$$

where $R_{1,2}$ is the distance of closest approach of an ion of mass m_1 and energy E_{inc} to characterize the interaction with the wall atom of mass m_2 in the elastic collision process. Taking these latter atoms to be of density N_2 then gives the relevant macroscopic cross-section Σ_{col} as

$$\Sigma_{col} = \sigma_{col} N_2 . \qquad (13.9)$$

Further, the mean-free-path of material penetration is

$$\lambda_{1,2} = \frac{1}{\Sigma_{col}} = \frac{1}{\pi R_{1,2}^2 N_2} . \qquad (13.10)$$

The above two parameters, $(\Delta E)_{max}$ and $\lambda_{1,2}$, can now be used to define a useful "sputtering ratio"

$$S = \frac{\text{Number of wall atoms ejected into the plasma}}{\text{Number of incident ions}} . \qquad (13.11)$$

A plausibility argument suggests that

$$S \propto (\Delta E)_{max} \qquad (13.12a)$$

and also that

$$S \propto \frac{1}{\lambda_{1,2}} . \qquad (13.12b)$$

Hence, it follows that

$$S = K \left[\frac{4m_1 m_2}{\lambda_{1,2}(m_1 + m_2)^2} \right] E_{inc} \propto E_{inc}\, \sigma_{col}\, N_2 \qquad (13.13)$$

where K is a proportionality constant and the latter expression emphasizes the particularly important dependencies.

Experimentally measured sputtering ratios are presented in Fig.13.6 for various first wall materials bombarded by monoenergetic deuterons. The sputtering ratios generally increase with energy of the incident ion–as suggested in Eq.(13.13)–until a point is reached where most collisions occur at such a depth that the probability of escape of the knock-ons is substantially reduced; thereafter, the yield decreases with incident energy. Hence, Eq.(13.13) is only applicable to the lower energy region of these curves since it neglects the effect of subsequent collisions or knock-ons.

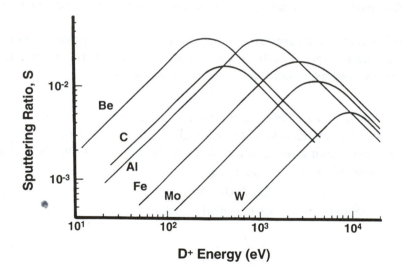

Fig. 13.6: Sputtering ratio of various materials as a function of incident deuterium ion energy.

In addition to physically weakening the wall as material is sputtered away, sputtered material that enters the plasma can have serious effects on plasma energetics. This will be considered next.

13.4   Wall Impurity Effects

The atoms sputtered from the first-wall into the plasma can lead to considerable degradation of the energy viability of the fusion reaction chain. The reason for this can be summarized as follows. As the sputtered impurity atoms enter the high temperature plasma, they collide with the plasma fuel ions and electrons and thereby become ionized. Only if these impurities feature a low proton number will they be completely ionized upon entering the plasma. For heavier impurity atoms (e.g. Fe, Ni, ...), full ionization will not be possible in the plasma, hence leaving them in states of incomplete ionization, notwithstanding the exhibition of high charge numbers depending on the plasma temperature. Evidently, the overall Z of the plasma is increased, and consequently also the bremsstrahlung radiation which was seen to vary with Z^2. Since several types of impurity ions may be present in a plasma at specific concentrations and different charge number, our previous formula for the bremsstrahlung power, Eq. (3.44), should, more correctly, be rewritten as

$$P_{br} = A_{br} \sum_i N_i Z_i^2 N_e \sqrt{kT_e} = A_{br} Z_{eff} N_e^2 \sqrt{kT_e} \qquad (13.14a)$$

where the plasma's effective charge number

$$Z_{eff} = \frac{\sum_i N_i Z_i^2}{N_e}, \quad i = \left\{ \begin{array}{l} \text{all ions distinguishable by} \\ \text{Z_i (including impurities)} \end{array} \right\} \qquad (13.14b)$$

has been introduced.

Additionally, a further important radiation process occurs which is due to the incomplete ionization of impurity atoms. Partly ionized atoms can be collisionally excited to a higher atomic energy level, and subsequently emit electromagnetic radiation by spontaneous transition to a lower energy level. The radiation frequency associated herewith is characteristic for each possible specific transition, and hence this radiation is called line radiation. In the case of a non-negligible impurity concentration in the plasma, the associated line radiation is seen to significantly contribute to the plasma radiation losses.

Since the line radiation P_{line} is proportional to both N_i and N_e, we introduce an ion-specific radiation parameter $\psi_{rad,i}$ by

$$\psi_{rad,i} = \frac{(P_{br})_i + (P_{line})_i}{N_i N_e} \quad \left[\text{Wm}^3 \right] \qquad (13.15)$$

which is displayed in Fig.13.7 as a function of plasma temperature. If this parameter is multiplied by the product of the electron density and the density of the i-th impurity, the respective radiation power due to both bremsstrahlung and line transitions is immediately found. Figure 13.7 now makes evident the substantial increase in radiation losses when high Z impurities enter the plasma. Only a very small concentration of such impurities can therefore be permitted in a fusion plasma if it is to be ignited. The straight dotted lines in Fig.13.7 indicate the normalized bremsstrahlung radiation for the case that the atoms of the considered element were fully stripped of electrons. As the plasma temperature increases, such states of complete ionization become more probable and the radiation parameters $\psi_{rad,i}$ are seen to asymptotically approach the respective corresponding straight dotted lines characterizing the bremsstrahlung from fully

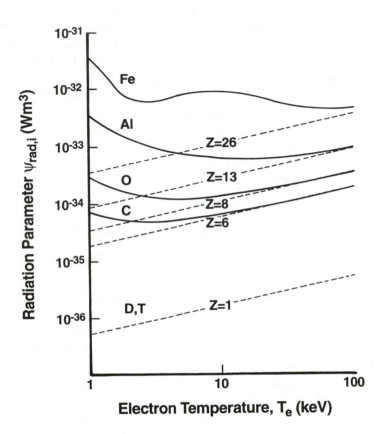

Fig. 13.7: Normalized radiation power, $\psi_{rad} = (P_{br} + P_{line}) / (N_i N_e)$, as it depends on the plasma electron temperature. The dotted lines denote the normalized bremsstrahlung radiation of the completely ionized respective atoms.

ionized atoms.

While a detailed analysis of the adverse effect of the impurity atoms may involve complex transport calculations, an appreciation of some of the implications can be formulated as follows. Consider an initially pure hydrogen-state unit volume of plasma containing deuterons and tritons of equal density and their associated electrons all at temperature T_1. That is, we have $N_d = N_t$, $N_i = N_d + N_t$, $N_i = N_e$ and a total thermal energy density in the fusion domain of

$$E_{th,1} = \tfrac{3}{2}(N_i + N_e)kT_1 .$$ (13.16a)

The contaminated state subsequently contains an additional number of N_z impurity ions and $N_z Z$ impurity electrons per unit volume; here Z is the ionization number for the atoms sputtered from the first wall. At thermal equilibrium with a temperature T_2 we therefore have

$$E_{th,2} = \tfrac{3}{2}(N_i + N_z + N_e + N_z Z)kT_2 . \tag{13.16b}$$

In the absence of energy injection and the imposition of sharing of thermal energy among all ions and electrons, we may take $E_{th,1} = E_{th,2}$ and obtain

$$T_2 = \frac{N_i + N_e}{N_i + N_e + N_z + N_z Z} T_1 \tag{13.17}$$

for which, evidently, $T_2 < T_1$. This plasma cooling effect may be more compactly represented by the introduction of an impurity ratio g_z defined by

$$g_z = \frac{N_z}{N_i} \tag{13.18}$$

so that Eq.(13.17) becomes, with $N_i = N_e$,

$$T_2 = \frac{1}{1 + \left(\dfrac{g_z(1+Z)}{2}\right)} T_1 . \tag{13.19}$$

The associated increased bremsstrahlung radiation power losses can be assessed by taking the ratio

$$\frac{P_{br,2}}{P_{br,1}} = \frac{A_{br}(N_e + N_z Z)^2 Z_{eff} \sqrt{kT_2}}{A_{br} N_e N_i \sqrt{kT_1}} \tag{13.20a}$$

with Z_{eff} according to Eq.(13.14b), however noting that the electron density here is $N_e + N_z Z$. Using the definition for the impurity ratio g_z of Eq.(13.18) then gives

$$\frac{P_{br,2}}{P_{br,1}} = (1 + g_z Z)^2 Z_{eff} \sqrt{\frac{T_2}{T_1}} \tag{13.20b}$$

and introducing the temperature ratio of Eq.(13.19) then yields

$$\frac{P_{br,2}}{P_{br,1}} = \frac{(1 + g_z Z)^2}{\sqrt{1 + \dfrac{g_z(1+Z)}{2}}} Z_{eff} \tag{13.20c}$$

which shows that bremsstrahlung power losses have indeed increased, even though the temperature was somewhat reduced ($T_2 < T_1$). Hence, this radiation emission, as well as the line radiation previously discussed and which may substantially contribute to plasma power losses, will further cool down the plasma.

Note that those fusion products having $Z > 1$ (helium) will result in greater bremsstrahlung radiation as well. Therefore, these so-called ash ions should be controlled in a fusion plasma so as not to accumulate to concentrations that are too high. The adverse impurity effect can be controlled by several methods. Two of the most important are the following: i) the use of divertors which involves magnetic field lines leading out of the plasma chamber to ion collectors so that ions following these lines do not hit the wall and hence do not produce impurities, and ii) the use of low-Z first-wall materials which do not cause such a large increase in the radiation loss even if their concentration in the plasma remains high.

13.5   ICF Chamber Protection

The selection of a method to protect an ICF chamber wall depends on a number of factors including the energy yield per pulse, the pulse repetition rate, and chamber vacuum requirements. In general, as the pulse rate increases, more protection is needed–such as thick liquid metal wall chambers; relatively large radius dry wall chambers could be considered for pulse energies less than 200 MJ.

Laser and heavy ion beam drivers require a high vacuum for good beam transmission; for example, lasers are limited to less than a few Torr (1 Torr $\approx$ 133.3 Pa) to prevent excessive scattering. In cases such as light ion beams, a higher pressure of the order of 5 to 50 Torr may be acceptable. If higher pressures are allowed, additional protection can be obtained by using a gas-fill in the chamber. The gas absorbs much of the radiation energy which is re-emitted and hits the chamber wall; this process significantly spreads out the time-width of the energy pulse thus reducing the shock effect.

Another important consideration for chambers with wetted walls, is the time required to clean–that is to purge–the chamber between pulses. This sets a limit for a maximum pulse rate. For example, it appears that about 1 s will be required to pump vapourized lithium out through an exhaust nozzle for a lithium-waterfall system. For such designs, a practical operating region could be a 1 Hz repetition rate with 100 MJ micro-explosions giving an average power of ~ 100 MW per chamber. Multiple chambers employing sequential switching of the laser beam from one to another might be used with a laser operating at a higher pulse rate.

13.6   μDT Channel

The first wall for a muon catalyzed fusion reaction will need to withstand pressures of the order of 10^8 Pa at moderate temperatures of about 10^3 K. Hence, this wall may be viewed as a thick cladding not unlike pressure tubes used in some fission reactors. However, while the design concept suggests some simplifications, significant problems remain to be studied; among these are the material requirements and the high pressure operation under conditions of high neutron influence.

13.7 Blanket Neutronics and Energetics

From a neutronic point of view, the blanket must be designed to provide an adequate tritium breeding ratio to sustain a substantial volumetric power density, allowing for continuous power extraction via heat exchange with the coolant, and to serve as shielding for equipment, especially for the superconducting magnet coils in MCF, and personnel.

Tritium is produced by the neutron-induced reactions

$$^6Li + n \rightarrow t + \alpha + 4.78 \text{ MeV} \qquad (13.21a)$$

and

$$^7Li + n \rightarrow t + \alpha + n - 2.47 \text{ MeV} \qquad (13.21b)$$

with a tritium breeding ratio C_t defined by

$$C_t = \frac{\text{total tritium production rate in blanket}}{\text{tritium destruction rate in core}}$$

$$= \frac{\int\limits_{V_b}\int\limits_{v_n} \sigma_{n6}(v_n) N_6 N_n(v_n) v_n dv_n d^3r + \int\limits_{V_b}\int\limits_{v_n} \sigma_{n7}(v_n) N_7 N_n(v_n) v_n dv_n d^3r}{\int\limits_{V_c} N_d N_t <\sigma v>_{dt} d^3r},$$

$$(13.22)$$

associated with isotropically distributed neutrons, $N_n(v_n)$, colliding with lithium nuclei at rest. Here N_6 and N_7 are the ^6Li and ^7Li atom densities in the blanket volume V_b, and σ_{n6} and σ_{n7} are their corresponding microscopic neutron absorption cross sections; N_n is the speed dependent density of neutrons in the blanket, v_n is the neutron speed and V_c is the fusion core volume. We point out that the n(^7Li, n, α)t reaction, Eq.(13.21b), is a threshold reaction and requires an incident neutron energy in excess of 2.47 MeV. For that, the neutron flux ($N_n v_n$) has to be considered in the fast energy range, that is the second integral in the numerator of Eq.(13.22) yields zero in the thermal and epithermal neutron energy range.

In the absence of tritium from other sources, it is necessary to have $C_t > 1$ in order to compensate for tritium transport losses during extraction and transfer as well as for its decay before injection into the fusion core. Indications for breeding capabilities can be obtained from upper limit estimates given in Table 13.1 for various blanket materials; as shown, adequate tritium breeding may be obtained if a neutron multiplier such as ^9Be or Pb is added.

Early blanket concepts employed liquid lithium as a coolant, thereby providing adequate tritium breeding. However, other considerations like pumping power requirements, the effect of magnetic fields on a flowing metal, and materials compatibility forced the development of alternative designs with lithium added in other forms. This includes various solid lithium compounds, molten salt fluids (e.g. 2LiF + BeF, called FLIBE), and a lithium-lead eutectic,

Material	Estimated Upper Limit Breeding Ratio, C_t
^{6}Li	1.1
Natural Li	0.9
^{9}Be + ^{6}Li (5%)	2.7
Pb + ^{6}Li (5%)	1.7

Table 13.1: Tritium breeding ratios for various materials, encompassing the entire fusion core.

17-Li 83-Pb. Table 13.2 summarizes calculated breeding ratios obtainable for a variety of materials in a "typical" blanket 1 cm thick with 10% volume fraction of 316 stainless steel, preceded by a 1 cm steel front-wall and backed by a 100 cm thick shield. Only the metallic lithium, LiO_2, and two of the Li-Pb eutectics appear to offer adequate tritium breeding. Consequently, use of the various solid breeders generally requires an added neutron multiplier.

Material	Calculated Tritium Breeding Ratio, C_t
17-Li 83-Pb	1.6
LiPb	1.4
FLIBE	1.1
$LiAlO_2$	0.9
LiO_2	1.3
Li_2SiO_3	0.9
Li_2ZrO_3	1.0

Table 13.2: Tritium breeding attainable with typical lithium bearing materials.

Since the blanket is exposed to high energy neutrons entering from the fusing plasma, the neutron density is a maximum in the first wall domain and then attenuates rapidly, even if a reflector zone completes the blanket composition. A consequence of this is that energy deposition will similarly vary with the depth of blanket penetration, Fig.13.8. The general trend of an exponential fall-off from the plasma side to the blanket interior must be considered in designing the coolant flow pattern and also in calculations of breeding, radiation damage, and activation. A maximum power density of ~ 80 Wcm^{-3} occurs in the multiplier zone, while the average is ~ 15 Wcm^{-3}. For comparison, power densities of ≤ 100 Wcm^{-3} apply to light and heavy water fission reactors. The fusion blanket region should operate at a high average temperature (≥ 1000 K) in order to facilitate a reasonable thermodynamic conversion efficiency.

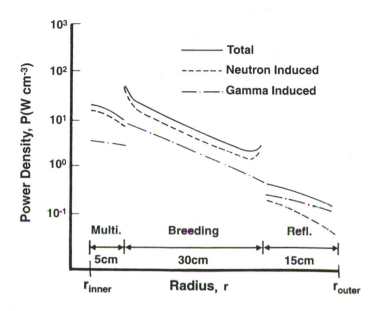

Fig. 13.8: Power density for a typical blanket using a beryllium neutron multiplier zone followed by a high concentration of LiAlO$_2$ and some H$_2$O with an outer graphite reflector.

Safety aspects suggest considering helium as an appropriate coolant since it is an inert gas which reacts with neither the lithium, the beryllium neutron multiplier nor other structural material. Further, it offers the advantage that the bred tritium is conveniently transported out of the blanket with the helium coolant flow.

In assessing the energetic performance of a fusion reactor blanket, we refer to the internal energy flows illustrated in Fig.13.9, from where it is evident that the energy removable from the blanket is

$$E_b^* = b\left(f_n E_{fu}^* + E_{rad}^* \right) + \sum_\ell E_{n\ell}^* \qquad (13.23)$$

where a blanket coverage factor b depending on the specific blanket geometry is introduced, since a fusion reactor blanket will feature several channels through its structure, e.g. for injection tubes, diagnostic equipment, etc., and hence cannot completely envelope the fusion plasma. Further, in Eq.(13.23), $E_{n\ell}^*$ accounts for the total energy released by an ℓ-type neutron-induced reaction. If exothermic, these reactions can then provide for multiplication of the energy of fusion neutrons having initially entered the blanket. To generalize such energy enhancement, it is convenient to account for it by the explicit blanket multiplication factor

$$M_b = \frac{b f_n E_{fu}^* + \sum_\ell E_{n\ell}^*}{f_n E_{fu}^*} \qquad (13.24)$$

allowing us to now rewrite Eq.(13.23) in the following form:

$$E_b^* = M_b f_n E_{fu}^* + b E_{rad}^* . \qquad (13.25)$$

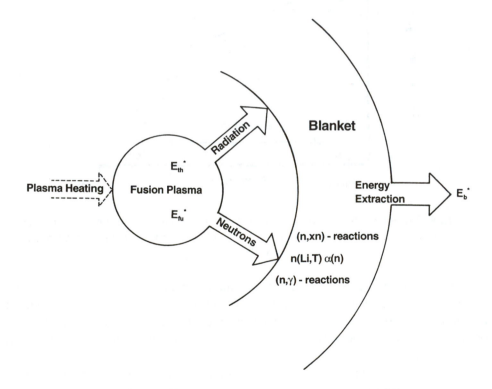

Fig. 13.9: Energy flows into and from a fusion reactor blanket.

Introducing specific power expressions based on reaction rate densities and the corresponding reaction Q-values, we find for Eq.(13.24)

$$M_b = b + \frac{\sum_\ell \int_{V_b} R_{n\ell} Q_{n\ell} d^3 r}{f_n \int_{V_c} R_{fu} Q_{fu} d^3 r} \tag{13.26a}$$

or, respectively, for d-t fusion and assuming lithium as the only neutron reactive substance in the blanket, we obtain

$$M_b = b + \frac{Q_{n6} \int_{V_b} \int_{v_n} \sigma_{n6} N_6 N_n (v_n) v_n dv_n d^3 r + Q_{n7} \int_{V_b} \int_{v_n} \sigma_{n7} N_7 N_n (v_n) v_n dv_n d^3 r}{f_{n,dt} Q_{dt} \int_{V_c} N_d N_t <\sigma v>_{dt} d^3 r}.$$

$$\tag{13.26b}$$

With the blanket composition and dimension known, and for a specified fusion plasma, M_b can be readily calculated and is seen to range from 1.3 to 1.8 for pure fusion blankets–those which do not contain fissionable material.

13.8   Radioactivation

In addition to normal structural considerations, it is necessary to evaluate neutron-induced radiation damage and radioactivation effects in the selection of both structural components and coolants. As discussed previously, radiation damage occurs by atom displacement and by nuclear transmutation involving primarily those producing ^{4}He; as expected the damage is most severe in the first-wall and associated structures on the side facing the fusion plasma. Atomic displacement rates and gas production rates are summarized in Table 13.3 for various materials placed in a neutron flux typical of the first-wall in a tokamak with a 1 MWm^{-2} neutron wall loading. The displacement rate is not strongly dependent on the type of material whereas the gas production rate is very sensitive to material choices. Nickel bearing alloys generally have a large ratio of gas-production and displacement rate. Lithium also possess a significant gas-production capacity but if this occurs in the liquid, pressure buildup and swelling are not a problem as it can be in solids. Neutron-induced   transmutations   in blanket materials also result in radioactivation which is most important with respect to reactor maintenance, and storage of reactor components. The level of radioactivation, along with other radioactivity aspects such as the tritium inventory, will be a key factor in determining the environmental impact of fusion reactors.

An illustration of the residual radioactivity of selected materials after a 2-year exposure is shown in Fig.13.10. The large variation in radioactive level and its effects with time for the various materials is a notable characteristic that must be considered. Inertial confinement fusion blanket designs using a thick liquid-metal

Material	Displaced atoms (10^7 atoms / s)	Helium production (10^7 atoms / s)	Hydrogen production (10^7 atoms / s)
Fe	3.6	35	150
Ni	3.9	130	400
Mn	3.6	27	100
Nb	2.3	9	30
Ti	5.0	34	50
Cu	4.9	32	170
^{6}Li		3100	3100
^{7}Li		360	370

Table 13.3: Typical atomic displacement and gas production for 1 MWm^{-2} first wall loading.

first-wall have a built-in advantage relative to minimizing activation of the chamber wall and structure. The falling liquid can reduce the neutron flux hitting this structure so that radioactivity levels are lowered by an order of magnitude or more relative to a dry wall.

Problems

13.1   Consider a neutron wall loading limit of 5 MWm^{-2} in a torus configuration with minor radius a = 2 m. If the plasma fuel ion densities are given by $N_d(r) = N_t(r) = N(r)$ of Eq. (6.54) where $N(0) = 10^{20}$ m^{-3}, what is the highest plasma temperature allowed?

13.2   Using the sputtering data for D$^+$ at 100 eV bombarding Fe in Fig.13.6, evaluate K in Eq.(13.13). Sketch the sputtering-curve predicted by this equation, and discuss any differences with Fig.13.6. Compute the time it would require for a flux of 10^6 deuterons·cm^{-2}·s^{-1} at 100 eV to sputter away 10% of the thickness of a 1 cm iron wall.

13.3   Consider a deuterium plasma at T_e = 10 keV containing 1% oxygen. Estimate the emitted radiation power using a weighted sum of the powers from the individual species.

13.4   Estimate the percentage of iron impurity in a d-t plasma that would cause the ideal ignition temperature to double.

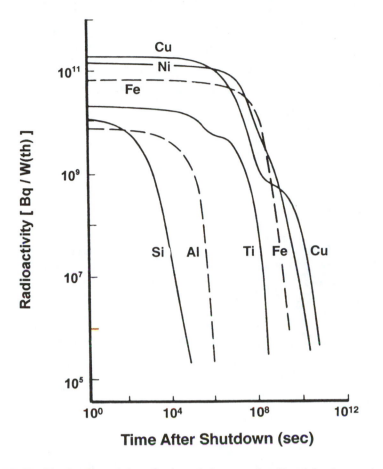

Fig. 13.10: Residual radioactivity of selected elements irradiated for 2 years in a typical first-wall flux of 1.5 MWm^{-2}.

13.5   Evaluate impurity effects on the Lawson criterion.

13.6   Consider a 50:50% MCF device having $T_i \approx T_e$, a plasma beta value of 0.2, $\psi \approx 10^{-3}$ and a doubly charged-ion impurity concentration of 1% of N_i. What is the modified ignition temperature (see problem 5.9)?

14. Tritium Fuel Dynamics

Tritium is of interest for nuclear fusion because, along with deuterium, it is a basic fuel for the most readily achievable fusion reaction. Since tritium does not occur naturally in sufficient quantities and is radioactive, its production and management creates some special dynamical features of relevance.

14.1 Tritium Properties

An essential role of a tritium nucleus is as a reactant in the fusion reaction

$$d + t \rightarrow n + \alpha , \qquad Q_{dt} = 17.6 \text{ MeV} . \tag{14.1}$$

Further, a triton is a radioactive beta emitter according to

$$t \rightarrow h + \beta^- , \qquad \lambda_t = 1.78 \times 10^{-9} \text{ s}^{-1}, \qquad \tau_{1/2} = 12.3 \text{ years} . \tag{14.2}$$

As previously used, h again represents the helium-3 nucleus; the maximum β^- energy is 18 keV with an average of ~5 keV. This property of nuclear instability is responsible for two important characteristics of tritium: it is naturally scarce and where it does exist, it is a radioactive hazard.

This natural scarcity of tritium, with the only known inventory being some 20 kg in the oceans and atmosphere where it is produced by reactions initiated by cosmic radiation–combined with the recognition that kilogram quantities of tritium may be required for a commercial central station fusion reactor–means that tritium will need to be bred on a substantial scale. As indicated in Sec.1.4 and Ch. 13, tritium breeding can occur by incidental neutron capture in the (heavy) water of fission reactors or by neutron capture in lithium in the blanket of a fusion reactor. The feature that every d-t fusion reaction produces one neutron, Eq. (14.1), means that the breeding by neutron induced reactions of one triton for every one destroyed would only be possible if no neutrons escaped or were lost to parasitic reactions. Since some neutron losses are unavoidable, a means of neutron multiplication is also essential for the breeding process.

An indication of the radiation hazard associated with tritium is suggested by calculating the decay rate of, say, 1 kg of tritium. From the definition of nuclear activity, Act, we have

$$Act = \left| \frac{dN_t^*}{dt} \right| = \lambda_t N_t^* , \tag{14.3}$$

where λ_t is the decay constant, Eq.(14.2). The total number of tritons N_t^*

associated with a given mass of tritium M_t is given by

$$M_t = N_t^* m_t \, , \tag{14.4}$$

where m_t is the mass of one tritium atom. Thus, the radioactive decay rate of 1 kg of tritium in units of disintegrations per second (dps) is given by

$$Act \, (1 \text{ kg of tritium}) = \frac{\lambda_t M_t}{m_t}$$

$$\tag{14.5}$$

$$= \frac{1.78 \times 10^{-9} \times 1}{5 \times 10^{-27}} = 3.56 \times 10^{17} \text{ s}^{-1} \, .$$

Translating this quantity into Curies, knowing that 1 Ci = 3.7×10^{10} dps (= 3.7×10^{10} Bq), the activity of 1 kg of tritium is equal to 10^7 Ci. To place this quantity into context, we add that only about 10^{-3} Ci of tritium activity can be handled without special licensing provisions. Evidently then, extreme care needs to be exercised in the management of large quantities of tritium.

Tritium, being a hydrogen isotope, can be readily transported by gaseous, liquid, and solid carriers. Its extraction is possible by catalytic exchange and cryogenic distillation processes. Two tritium transport characteristics have been found useful. For the case of a local tritium density gradient ∇N_t in a diffusion medium, Fick's rule of diffusion provides for a tritium current $\mathbf{J}_t$ given by

$$\mathbf{J}_t = -D_t \nabla N_t \tag{14.6}$$

where D_t is the tritium diffusion coefficient. Also, tritium permeation through a barrier containing a differential tritium density yields a tritium flux ϕ_t well represented by

$$\phi_t \propto K_t \left(\frac{\sqrt{N_{t,1}} - \sqrt{N_{t,2}}}{Z} \right) . \tag{14.7}$$

Here $N_{t,1}$ and $N_{t,2}$ are the upstream and downstream tritium densities, Z is the barrier thickness, and K_t is the tritium permeation constant. Parameters such as D_t and K_t are strongly material and temperature dependent, and, additionally, the permeation constant is a function of surface conditions. Low temperature, multiple wall barriers appear to be the most promising means of tritium containment.

14.2   Continuous D-T Burn

Tritium fuel cycle characteristics are largely determined by the mode of reactor operation and the source of tritium. We consider first a d-t fusion reactor operating at constant power for which tritium is supplied externally. Hence, the fuel leaks out of the containment vessel along with the fusion reaction products as suggested in Fig.14.1. Our interest is now in the specification of the tritium inventory in the fusion core and the tritium inflow requirements in terms of some basic reactor operational features.

Fusion Core

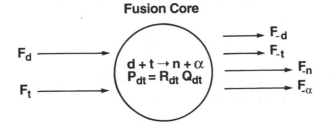

Fig. 14.1: Flow of fuel and reaction products for the continuous burn d-t fusion reactor.

The starting point is the steady state fusion reaction rate density

$$R_{dt} = N_{d,c} N_{t,c} <\sigma v>_{dt} \tag{14.8}$$

where $N_{d,c}$ and $N_{t,c}$ are the number of deuterons and tritons in the unit-volume fusion reaction chamber. The associated constant fusion power density is evidently

$$P_{dt} = R_{dt} Q_{dt} = N_{d,c} N_{t,c} <\sigma v>_{dt} Q_{dt} \tag{14.9}$$

with d-d reactions neglected. Consider next the fuel inflow-outflow-destruction processes for the fusion core so that

$$\frac{dN_{d,c}}{dt} = F_d - F_{-d} - R_{dt} = f_d F_d - R_{dt} \tag{14.10}$$

and

$$\frac{dN_{t,c}}{dt} = F_t - F_{-t} - R_{dt} = f_t F_t - R_{dt} . \tag{14.11}$$

Here $F_{()}$ are the fuel inflow-outflow rates depicted in Fig.14.1 and f_d and f_t are the respective burn fractions of deuterium and tritium; tritium decay is here justifiably neglected since the length of time the tritium is in transit is much less than its half-life.

Since an MCF plasma fusion core is optically thin for neutrons, all the fusion neutrons will leak therefrom and hence $F_{-n} = R_{dt}$. The alphas produced in fusion reactions collisionally interact with the plasma ions and electrons, and– attributable to their electric charge–they are more or less well confined to the respective magnetic field configuration. Their outflow rate $F_{-\alpha}$ is managed by external means, e.g. by divertor extraction. Neglecting side reactions, deuterons and tritons are burned at equal rates yielding

$$f_t = f_d \quad \text{for} \quad F_d = F_t . \tag{14.12}$$

A connection between the fusion core tritium inventory and the power level follows from Eq.(14.9):

$$N_{t,c} = \frac{P_{dt}}{N_{d,c} <\sigma v>_{dt} Q_{dt}} . \tag{14.13}$$

Thus, the operation of the fusion core which maximizes $<\sigma v>_{dt}$ also minimizes the tritium core inventory.

The tritium injection requirements for steady-state operation, $dN_{t,c}/dt = 0$, are given by Eq.(14.11):

$$F_t = \frac{R_{dt}}{f_t} = \frac{P_{dt}}{f_t Q_{dt}}$$ (14.14)

providing therefore, a rigid relationship between the variables of power, burn fraction, and tritium injection rate.

14.3   Pulsed D-T Burn

We next consider a pulsed d-t fusion reactor for which, as for the previous case, tritium is supplied externally and ideal confinement in the fusion core exists. This pulsed burn mode can well be characterized by three stages of tritium injection, tritium burn, and tritium purging, Fig.14.2. During injection and purging, some of the tritium will be lost by particle transport into the containment components while during the burn, tritium is destroyed by the fusion process. It is the burn stage that is of most interest to us.

As for the preceding continuous-burn model, we begin with a dynamical description during the burn time τ_b. A rapid pulse implies a rapid change in plasma temperature so that sigma-v during the burn cannot be taken to be a constant and we will therefore represent this time dependence by $<\sigma v(t)>_{dt}$. The consequence of this is that the fuel inventory will vary similarly with time during the burn so that the time dependent reaction rate density R_{dt} is

$$R_{dt}(t) = N_{d,c}(t)N_{t,c}(t) < \sigma v(t) >_{dt} .$$ (14.15)

The dynamical equations for the fuel density in the core are simply

$$\frac{dN_{d,c}}{dt} = -R_{dt}(t) ,$$ (14.16)

and

$$\frac{dN_{t,c}}{dt} = -R_{dt}(t) .$$ (14.17)

Note, here, the absence of any inflow-outflow terms during the burn time. The instantaneous d-t power in a unit volume at any time in the burn interval is

$$P_{dt}(t) = R_{dt}(t)Q_{dt} = N_{d,c}(t)N_{t,c}(t) < \sigma v(t) >_{dt} Q_{dt} .$$ (14.18)

We cannot proceed further without a knowledge of the time dependence of $<\sigma v(t)>_{dt}$. Recalling the definition of this sigma-v parameter, Eq.(2.29) implies that $<\sigma v(t)>_{dt}$ could be determined if the time variation of the deuterium and tritium velocity distributions were known. This may, in principle, be obtained from a detailed time dependent analysis of how the pulse-injection energy is transformed into fuel kinetic energy together with other concurrent energy transformation processes. However, this analysis is both difficult and tedious, and falls outside the scope of our objectives here.

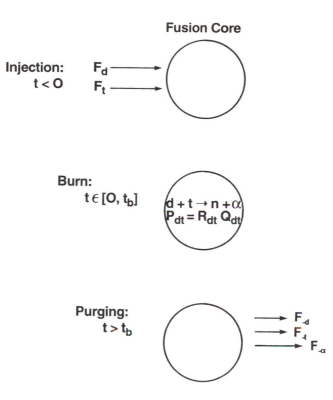

Fig. 14.2: Depiction of three stages associated with each reactor operating pulse.

Still, even without a detailed knowledge of the pulse energetics, we can outline the dominant time variations of the tritium fuel. Consider then, at the beginning of a typical burn cycle, an injected energy pulse spread over a short time period relative to the fusion power pulse. Ionization occurs promptly and the kinetic energy of the fuel ions rises rapidly to initiate fusion reactions and thus triton destruction by fusion. The alpha particle fusion products may thereupon continue to heat the ions to sustain fusion power production. Eventually, a variety of power losses–leakage, radiation etc.–will cool the plasma until the power pulse can be considered terminated. We suggest some of the time variations for a typical pulse in Fig.14.3.

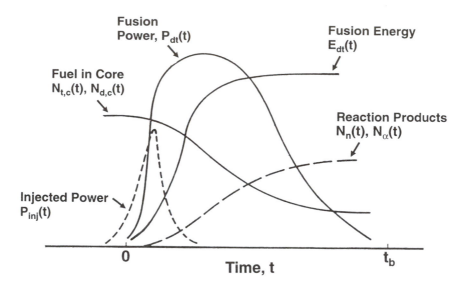

Fig. 14.3: Schematic of power-energy and fuel reactant variations during a d-t burn pulse.

14.4   Self-Sufficient Tritium Breeder

Consider now the case of a d-t fusion power device equipped with a tritium breeding blanket, the purpose of which is to breed tritium for the core. We will retain the steady-state operational mode description and also assume that the bred tritium is extracted from the blanket and supplied to the core on a time scale very short relative to the tritium decay half-life; hence, the decay of tritium need not be incorporated into the analysis. Figure 14.4 provides a schematic of the relevant particle flows and reaction rates.

The two important reaction rates are tritium destruction by fusion in the core

$$R_{dt}^* = \int_{V_c} N_{d,c} N_{t,c} <\sigma v>_{dt} d^3 r \qquad (14.19)$$

and tritium breeding in the blanket

$$R_{n\ell,t}^* = \int_{V_b} \int_{v_n} N_n(v_n) \cdot v_n \left[\sigma_{n6}(v_n) \cdot N_6 + \sigma_{n7}(v_n) \cdot N_7\right] dv_n d^3 r. \qquad (14.20)$$

Here V_c and V_b are the core and blanket volumes, N_n is the neutron density, v_n is the neutron speed, N_6 and N_7 are the lithium-6 and lithium-7 densities and σ_{n6} and σ_{n7} are the corresponding neutron absorption cross sections for tritium breeding.

Replacement of the burned tritium requires

$$R^*_{n\ell,t} \geq R^*_{dt} \, . \tag{14.21}$$

That is, the tritium production rate must be equal to or exceed the tritium destruction rate. In practice, tritium losses in the overall cycle by radioactive decay and transport into containment walls will occur so that for operational reasons $R^*_{n\ell,t} > R^*_{dt}$; this is incorporated in the tritium breeding ratio C_t by requiring

$$C_t = \frac{R^*_{n\ell,t}}{R^*_{dt}} > 1 \, . \tag{14.22}$$

We will next consider some dynamical aspect of the tritium inventory in the blanket which incorporates tritium decay and transport.

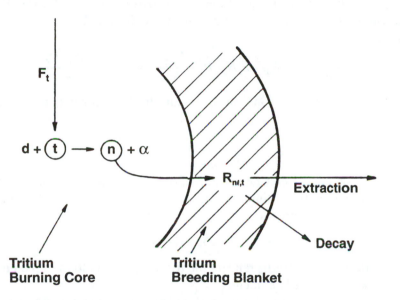

Fig. 14.4: Tritium burning, breeding, and other processes in a fusion reactor.

14.5   Tritium Dynamics in the Blanket

The history of tritium accumulation in the breeding blanket and its eventual extraction for recycle to the fusion reactor core is of paramount importance. Consider then a unit volume in the blanket in which tritium gains and losses occur:

$$\frac{dN_{t,b}}{dt} = R_{+t} - R_{-t} \, . \tag{14.23}$$

The gain-rate is clearly proportional to the rate of neutron production in the core so that, in view of the preceding section, we write

$$R_{+t} = R_{n\ell,t} = C_t R_{dt} \cdot \tag{14.24}$$

Tritium removal in a unit volume of the breeding blanket occurs by radioactive decay and extraction:

$$R_{-t} = \left(R_{-t,b}\right)_{decay} + \left(F_{-t,b}\right)_{extract} \cdot \tag{14.25}$$

Here, the radioactive decay component is

$$\left(R_{-t,b}\right)_{decay} = \lambda_t N_{t,b}(t) \tag{14.26}$$

and the extraction rate is taken to be a constant fraction of the tritium in the blanket:

$$\left(F_{-t,b}\right)_{extract} = \frac{N_{t,b}(t)}{\tau_{t,b}} \cdot \tag{14.27}$$

Here $\tau_{t,b}$ is the mean residence time of the tritium in the blanket. Hence, the dynamical equation for the tritium inventory in the blanket, Eq.(14.23), is therefore

$$\frac{dN_{t,b}}{dt} = C_t R_{dt} - \lambda_t N_{t,b}(t) - \frac{N_{t,b}(t)}{\tau_{t,b}} = C_t R_{dt} - \frac{N_{t,b}(t)}{\tau_b} \tag{14.28}$$

where an equivalent time constant for tritium in the blanket τ_b had been defined as

$$\frac{1}{\tau_b} = \lambda_t + \frac{1}{\tau_{t,b}} \cdot \tag{14.29}$$

With R_{dt}, C_t and τ_b as constant, Eq.(14.28) is in a standard form amenable to solution given an initial tritium density in the blanket. Suppose that start-up of a fusion reactor is of interest so that $N_{t,b}(0) = 0$, then the solution of Eq. (14.28) is

$$N_{t,b}(t) = C_t R_{dt} \tau_b \left(1 - e^{-t/\tau_b}\right) \cdot \tag{14.30}$$

The substitution $R_{dt} = P_{dt}/Q_{dt}$ renders the tritium accumulation in the unit volume of the blanket particularly evident.

The mean residence time of the tritium bred in the blanket, $\tau_{t,b}$ of Eq.(14.27), now appears as an important operational parameter. For cases where $\tau_{t,b}$ is short relative to the inverse tritium decay constant, i.e., $\tau_{t,b} \ll 1/\lambda_t$, or, alternatively, very long, i.e., $\tau_{t,b} \gg 1/\lambda_t$, a most distinct tritium accumulation and equilibrium concentration in the blanket results, Fig.14.5. The asymptotic quantity of accumulated tritium is thus directly proportional to the power and the time constant τ_b.

Some numerical experimentation with Eq.(14.30) will show that it may take decades for the bred tritium to approach the equilibrium value $N_{t,b}(\infty)$. It may also be desirable to introduce either continuous or periodic removal of the tritium while its accumulation is still small; we also illustrate this periodic batch removal of the tritium in Fig.14.5.

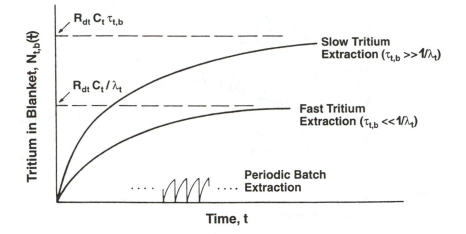

Fig. 14.5: Accumulation of tritium bred in the blanket for various extraction modes.

14.6 External Tritium Stockpile Dynamics

An actual fusion reactor will evidently be associated with an external tritium-fuel stockpile, much as a fission reactor needs to have an assured external supply of fissile fuel. We depict such a system in schematic form in Fig.14.6 where we also show the tritium flows and processes which will affect the inventory of tritium in the external stockpile.

Our preceding discussion involved the description of the tritium destruction rate in the fusion core as well as the tritium production rate and accumulation in the blanket. As suggested in Fig.14.6, tritium bred in the blanket may decay or it may be extracted for deposit in the external stockpile; once in the stockpile, tritium may also decay, it may also be withdrawn for burning in the fusion core, or it may be lost by transport processes such as diffusion. The dynamical equation for the tritium in the stockpile is therefore

$$\frac{dN_{t,x}^*}{dt} = F_{+t,x}^* - F_{-t,x}^* - \left(R_{t,x}^*\right)_{decay} - \left(R_{t,x}^*\right)_{loss} \tag{14.31}$$

where the $F_{()}$ terms are the flow rates suggested in Fig.14.6.

The tritium supply rate to the stockpile may well be taken identical to the tritium extraction rate from the blanket so that according to Eq.(14.27)

$$F_{+t,x}^* = \frac{N_{t,b}^*(t)}{\tau_{t,b}} \tag{14.32}$$

where $N_{t,b}^*(t)$ is the tritium inventory in the blanket at time t produced by neutron capture in lithium and $\tau_{t,b}$ is the mean residence of the bred tritium in the blanket.

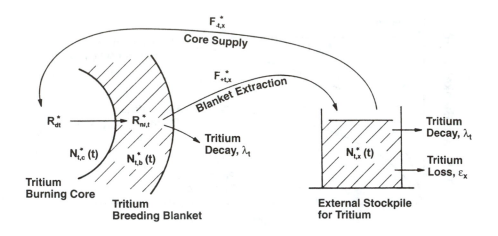

Fig. 14.6: Illustration showing various tritium flows.

The prompt removal rate of tritium from the stockpile should–for obvious reasons–be equal to the tritium destruction rate in the core:

$$F^*_{-t,x} = R^*_{dt} \, . \tag{14.33}$$

Further, radioactive decay losses from the external stockpile are

$$\left(R^*_{t,x}\right)_{decay} = \lambda_t \, N^*_{t,x}(t) \, . \tag{14.34}$$

Finally, for the present illustration, we assume that the various transport losses from the stockpile are a constant fraction of the inventory such that

$$\left(R^*_{t,x}\right)_{loss} = \varepsilon_x N^*_{t,x}(t) \, . \tag{14.35}$$

Substituting Eqs.(14.32) to (14.35) into Eq.(14.31) then specifies the dynamical state of the tritium inventory in the external stockpile as

$$\frac{dN^*_{t,x}}{dt} = \frac{N^*_{t,b}(t)}{\tau_{t,b}} - R^*_{dt} - \left(\lambda_t + \varepsilon_x\right)N^*_{t,x}(t) \, . \tag{14.36}$$

A solution of this equation requires an examination of the various terms and an imposition of some reactor operational modes.

With λ_t, ε_x, C_t, and $\tau_{t,b}$ as system constants and also taking R^*_{dt} to be constant, the time dependence of the tritium in the blanket $N^*_{t,b}(t)$ is given by Eq.(14.30). Under these conditions, Eq.(14.36) can be specified by the first order differential equation

$$\frac{dN^*_{t,x}}{dt} = \frac{C_t R^*_{dt} \tau_{t,b}}{\tau_{t,b}} \left(1 - e^{-t/\tau_{t,b}}\right) - R^*_{dt} - \left(\lambda_t + \varepsilon_x\right)N^*_{t,x}(t) \, , \; N^*_{t,x}(0) = N^*_{t,0} > 0 \, . \tag{14.37}$$

Here we have also shown an initial condition for the tritium in the stockpile at t = 0. This equation can be cast into the generic form

$$\frac{dN^*_{t,x}}{dt} = A_0 - A_1 e^{-t/\tau_b} - A_2 N^*_{t,x}(t) \tag{14.38}$$

where

$$A_0 = R^*_{dt}\left[C_t\left(\frac{\tau_b}{\tau_{t,b}}\right) - 1\right] \tag{14.39a}$$

$$A_1 = R^*_{dt}C_t\left(\frac{\tau_b}{\tau_{t,b}}\right) \tag{14.39b}$$

$$A_2 = \left(\lambda_t + \varepsilon_x\right). \tag{14.39c}$$

While an explicit solution for the time dependence of the amount of tritium in the stockpile, $N_{t,x}^*(t)$, can be obtained by solving the above inhomogeneous first-order differential equation, Eq.(14.38), we will find it more instructive to examine the differential equation itself; this will provide for a better understanding of how the various parameters influence the availability of this essential fuel.

Prior to start-up, the tritium in the external stockpile simply decays. Thus, at $t=0$, when the tritium inventory is $N_{t,x}^*(0) = N_{t,0}$, an instantaneous withdrawal from the stockpile to the core takes place and gradual replenishment from the blanket to the stockpile is also initiated; the differential equation thus describes the slope of $N_{t,x}^*(t)$ at any point in time.

Within a sufficiently short time after startup, e.g. $t = 0^+$, we find from Eq.(14.38) that

$$\text{Slope of } N^*_{t,x}(t)\Big|_{t=0^+} = A_0 - A_1 - A_2 N^*_{t,0} = -R^*_{dt} - \left(\lambda_t + \varepsilon_x\right)N^*_{t,0} \tag{14.40}$$

implying that the tritium inventory will initially decrease at a rate dependent upon the magnitude of the initial tritium inventory and fusion reaction rate.

Further, for t sufficiently large, the central term of Eq.(14.38) will vanish and hence

$$\text{Slope of } N^*_{t,x}(t)\Big|_{t\to\infty} = A_0 - A_2 N^*_{t,x}(\infty). \tag{14.41}$$

This then defines the asymptotic tritium inventory $N_{t,\infty}^*$ in the stockpile thus given by

$$N^*_{t,\infty} = \frac{A_0}{A_2} = R^*_{dt}\left[\frac{C_t\left(\frac{\tau_b}{\tau_{t,b}}\right) - 1}{\lambda_t + \varepsilon_x}\right], \tag{14.42}$$

with the general characteristic time variations of tritium suggested in Fig.14.7. Note that a minimum may exist unless $C_t(\tau_b / \tau_{t,b}) = 1$ in which case a zero tritium inventory will be attained.

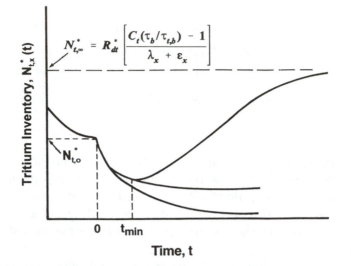

Fig. 14.7: Characteristic tritium stockpile inventory as a function of various system
parameters.

Problems

14.1   According to Eq.(14.2), an inventory of tritium will become depleted but
helium-3 (h) will be produced. This newly bred fuel could supply a fusion
reaction based on the $d + h \rightarrow p + \alpha$ cycle. If tritium is thus an energy "debit" due
to its decay, what is the corresponding energy "credit" for the production of
helium-3 as a function of time?

14.2   Compute the $N_t = N_d$ concentration in a fusion core generating 100 kW
per litre for various temperatures from 1 keV to 500 keV.

14.3   Determine an explicit solution for Eq. (14.37).

14.4   Examine a self-sufficient tritium breeding fusion reactor such that $N_{t,x} \rightarrow 0$
as $t \rightarrow \infty$.

14.5   How long, after start-up, until the tritium inventory in the external
stockpile attains a minimum?

15. Fusion-Fission Integration

A nuclear fusion core may be surrounded by a blanket in which the neutrons from the fusion reactions sustain tritium breeding as well as fissile fuel breeding and–to a varying extent–fission energy production. The fissile fuel bred in such a hybrid is to supply the fuel requirements for "client" fission reactors and, by the consequent energy credit generated, alleviate some energy balance constraints on the fusion component.

15.1   Conceptual Description

The potentially recoverable energy from a fusing domain consists of electromagnetic radiation, the kinetic energy of charged nuclear particles, and the kinetic energy of neutrons.

The neutron is of particular interest for the following reasons: while its kinetic energy can be utilized–that is converted into thermal energy as it slows down by nuclear collisions in the blanket–the neutron itself remains a valuable particle to induce selected nuclear reactions. As previously discussed, the tritium for fuel self-sufficient d-t fusion reactors must be bred by neutron capture in lithium. Hence, for d-t fusion, one essential use of the fusion neutron is to breed tritium in order to close the fuel cycle. The reactions and reaction linkages involved are

$$d + t \rightarrow n + \alpha$$
$$n + \begin{Bmatrix} {}^6Li \\ {}^7Li \end{Bmatrix} \rightarrow \begin{Bmatrix} 0 \\ n \end{Bmatrix} + \alpha + t \ . \tag{15.1}$$

Further thought suggests other productive uses of the 14.1 MeV fusion neutron. With its high kinetic energy, the fusion neutron has access to numerous (n,xn) neutron multiplying reactions, Fig.15.1, and these reactions can result in a number of neutrons in excess of what is required for tritium breeding. Consequently, the spare neutrons can be used for other purposes; alternatively, if a fuel cycle such as catalyzed-D described in Ch. 7 is used, tritium breeding is not required so that the entire neutron population is available for other purposes. An interesting system concept is to surround the fusion chamber with a blanket of fertile nuclear fuel, that is ${}^{238}U$ or ${}^{232}Th$, so that the neutrons can breed fissile fuel and/or aid in sustaining fission reactions in the intrinsically subcritical blanket.

Since each fission event yields ~200 MeV, the blanket serves the function of energy multiplication and can also serve as a "fuel factory" for fission reactors. This concept can be further clarified by considering three dominant classes of neutron reactions which, for illustrative purposes, are assumed to occur in separate regions of the blanket.

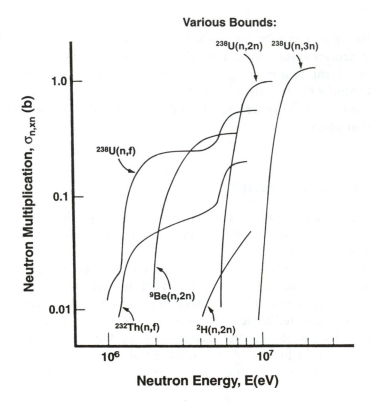

Fig. 15.1: Neutron multiplication cross sections accessible to neutrons from d-t fusion.

Immediately adjacent to the fusion core, a region is envisioned with a high concentration of isotopes possessing a significant (n,xn) neutron multiplication cross section (i.e., x > 1) as shown in Fig.15.1; the dominant reaction is therefore of the type

$$n + {}^{A}Z \rightarrow xn + {}^{A-x+1}Z, \quad x > 1 \tag{15.2}$$

where ${}^{A}Z$ represents a typical neutron multiplier. A location close to the fusion core capitalizes on the high energy of fusion neutrons to increase the neutron population by (n,xn) reactions which require neutron energies in excess of specific thresholds.

The next blanket layer is taken to contain fertile materials (${}^{232}Th$, ${}^{238}U$) which transmute into fissile materials (${}^{233}U$, ${}^{239}Pu$) by the processes

$$n + \left\{ \begin{array}{l} ^{232}Th \\ \\ ^{238}U \end{array} \right\} \xrightarrow{\gamma} \left\{ \begin{array}{l} ^{233}Th \\ \\ ^{239}U \end{array} \right\} \xrightarrow{\beta^-} \left\{ \begin{array}{l} ^{233}Pa \\ \\ ^{239}Np \end{array} \right\} \xrightarrow{\beta^-} \left\{ \begin{array}{l} ^{233}U \\ \\ ^{239}Pu \end{array} \right\} \quad (15.3a)$$

or, more simply

$$n + g \rightarrow f \quad (15.3b)$$

with g representing the fertile fuel and f denoting the fissile nuclei bred, some of which may be fissioned by neutron absorption and thereby generate significant amounts of energy.

The outer most blanket region is used for tritium breeding by neutron capture in lithium via

$$n + \left\{ \begin{array}{l} ^{6}Li \\ ^{7}Li \end{array} \right\} \rightarrow \alpha + t + \left\{ \begin{array}{l} 0 \\ n \end{array} \right\} \quad (15.4a)$$

or

$$n + \ell \rightarrow t, \quad (15.4b)$$

where ℓ denotes the lithium fuel.

It is thus evident that the hybrid blanket produces both energy and fissile fuel, and depending on the design objectives, one of these functions can be emphasized. The ratio of fissile fuel nuclei bred per unit fusion energy released is therefore an important parameter for the characterization of such a system. For example, in designs emphasizing the fuel factory approach where energy production is de-emphasized so the plant need not be a key contributor to an electrical network–and thereby also freeing it for a more flexible operating schedule–this ratio would be maximized. This may be accomplished by selecting materials in order to minimize fission reactions in the blanket while maximizing fissile breeding reactions.

These three dominant processes and the corresponding blanket domains are depicted in Fig.15.2. Other arrangements are possible, but the order chosen here for the various regions follows a functional pattern intended to make enhanced use of the fusion-source neutron energy: (i) neutron multiplication is most productively accomplished with high-energy neutrons and hence the blanket section designed for this process occurs close to the fusion core for immediate access to the 14.1 MeV fusion neutrons before they slow down; (ii) next, fissile fuel breeding is best accomplished with intermediate energy neutrons; (iii) finally, after the neutrons have slowed down, the 1/v-dependence of the neutron capture cross section of ^{6}Li ensures efficient tritium breeding.

15.2 Energy Multiplication

A detailed quantitative assessment of the energy multiplication capacity of a hybrid requires a specific blanket design followed by a detailed neutronic analysis. However, an indication of this important energy gain can be obtained by employing a simplified lumped parameter characterization of the several processes occurring in the blanket and associated reactors.

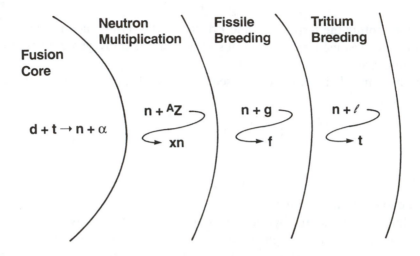

Fig. 15.2: Depiction of various neutron-nucleus reactions in the blanket surrounding a d-t fusion reactor.

We define C_b to be the average number of neutrons produced in the blanket by all neutron multiplication processes, expressed on a per fusion reaction basis. One of the neutrons so produced must be used for breeding tritium while the remaining C_b-1 neutrons could be used for fissile fuel breeding. Allowing for parasitic neutron captures and losses such as neutron leakage gives finally $\varepsilon_b(C_b-1)$ as the total number of fissile nuclei produced. Each of the bred fissile nuclei is able to eventually generate Q_{fi} units of energy, which will take place–if the bred fuel is extracted and transported to client power plants–in associated fission reactors. Then, the total nuclear energy generated by the hybrid fusion-fission system must also include the breeding capacity of the medium in which these eventual fissions take place. Letting C_{fi} be the conversion ratio of these associated fission reactors, a total fission energy of $Q_{fi}/(1-C_{fi})$ is therefore eventually generated and attributed to each initial fissile breeding reaction in the fusion reactor blanket. The total energy generated per unit initial fusion reaction energy release Q_{fu} defines an energy multiplication and is therefore given by

$$M_E = \frac{Q_{fu} + Q_{fi,total}}{Q_{fu}}$$

$$= \frac{Q_{fu} + \varepsilon_b (C_b - 1) Q_{fi} / (1 - C_{fi})}{Q_{fu}} \tag{15.5}$$

$$= 1 + \varepsilon_b \left(\frac{C_b - 1}{1 - C_{fi}} \right) \left(\frac{Q_{fi}}{Q_{fu}} \right), \quad C_{fi} < 1, \quad C_b > 1.$$

Here M_E may well be called the energy multiplication capacity and is displayed in Fig.15.3. Clearly, the energy multiplication of a fusion reactor operating in tandem with a fission reactor can be substantial compared to that for a stand-alone fusion reactor ($M_E = 1$).

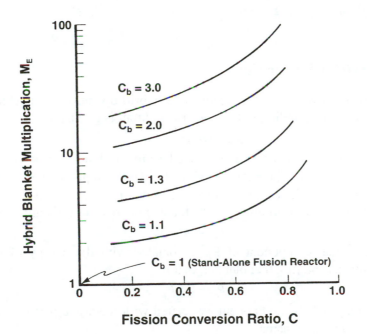

Fig. 15.3: Energy multiplication capacity of a d-t fusion-fission hybrid reactor assuming the neutron availability-for-breeding factor $\varepsilon_b \approx 0.8$.

The fundamental reason for the remarkable energy multiplication of a fusion hybrid reactor operating in tandem with a fission system lies in the complementary nature of fusion and fission reactions. A d-t fusion reaction results in one neutron and a total energy release of 17.6 MeV while a fission reaction results in 2 to 3 neutrons and ~200 MeV. Consequently, fission reactions can be viewed as energy "rich" and fusion reactions, by this yardstick, as energy "poor". Further, a hybrid fusion reactor blanket may well regenerate the fuel it consumes, i.e. breed tritium and have some neutrons left over. In contrast, improving the performance of a fission breeder reactor–which produces abundant amounts of energy–demands an enhanced neutron population. Hence, a "neutron-for-energy" and "energy-for-neutron" exchange in the hybrid provides mutual advantages.

Another way of viewing fusion-fission energetics is to note that a neutron used for tritium breeding in support of a fusion reaction contributes basically 17.6 MeV of energy while a neutron used to support fission energy by fissile fuel breeding adds more than 10 times as much energy to the entire yield of the system. Thus, the high energy neutrons from a fusion reaction possess a greater capacity for neutron multiplication than thermal neutrons in a thermal reactor, reinforcing the "neutron rich / energy poor" view of fusion. This contrasts to the recognition that on a basis of "energy per initial fuel mass" involved, fusion is exceedingly more "energy rich".

15.3   Hybrid Power Flow

The dominant power flow for steady-state hybrid operation is suggested in Fig. 15.4 and the corresponding station electrical output is given by

$$P_{net}^* = \eta_b \, P_{b,t}^* - P_{in}^* \, . \tag{15.6}$$

Here $P_{b,t}^*$ is the thermal power extracted from the blanket and converted into electrical form with efficiency η_b while P_{in}^* is the input power supplied with efficiency η_{in} to the device in order to sustain the fusion reactions.

Using an obvious definition for the blanket energy multiplication, we write

$$P_{b,t}^* = M_b \, P_n^* \tag{15.7}$$

where P_n^* is that component of P_{dt}^* associated with the energy of the fusion-source neutrons entering that blanket; that is

$$P_n^* = 0.8 P_{dt}^* \, . \tag{15.8}$$

The remaining $0.2 \, P_{dt}^*$ component associated with the alpha particles is assumed to be retained in the plasma.

Next, we define an effective fusion-component input power multiplication by

$$M_{fu,e} = \frac{\eta_b \, P_n^*}{P_{in}^*} \tag{15.9}$$

giving for the total station power, Eq.(15.6),

$$P_{net}^* = \eta_b \, P_n^* \, M_b - \frac{\eta_b \, P_n^*}{M_{fu,e}} = \eta_b \, P_n^* \left(M_b - \frac{1}{M_{fu,e}} \right) . \tag{15.10}$$

Note that upon introduction of the fusion plasma Q-value, Eq. (8.6), which here for steady state is

$$Q_p = \frac{P_{dt}^*}{\eta_{in} P_{in}^*} = \frac{P_n^*}{0.8\eta_{in} P_{in}^*} \ , \tag{15.11}$$

the effective fusion-component input power multiplication is represented by

$$M_{fu,e} = 0.8\,\eta_{in}\,\eta_b\,Q_p \ . \tag{15.12}$$

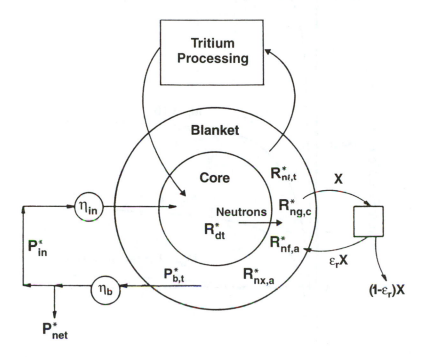

Fig. 15.4: Power and fuel flow in a fusion-fission hybrid; the symbols $R_{()}$ identify reaction rates.

The significant result here is that the condition for energy viability of a hybrid is now

$$M_b\, M_{fu,e} > 1 \tag{15.13}$$

and contrasts to a stand-alone fusion reactor for which, evidently, it is necessary to have $M_{fu,e} > 1$. Since the attainment of $M_{fu,e} > 1$ is a demanding technical problem and since $M_b > 1$ seems readily possible, the hybrid would have the possible advantage that some critical plasma parameters could be relaxed when compared to a pure fusion system. This could be an important motivation for hybrid operation in the early development of fusion systems.

The important consequences of Eq.(15.10) for the hybrid are depicted in Fig.15.5. Note that for $M_b > 1$, the energy multiplication of the fusion reactor, that is $M_{fu,e}$, can be less than unity and still provide for an energetically viable overall system.

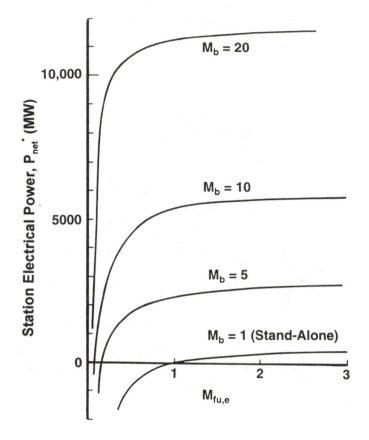

Fig. 15.5: Fusion-fission hybrid electrical power as a function of the fusion-component input energy multiplication ($M_{fu,e}$) and the blanket energy multiplication (M_b). Here, $P_{dt}^{*} =$ 2500 MW and $\eta_b = 0.3$.

15.4 Hybrid Fuel Breeding

The preceding discussion stressed the energy multiplication potential of the fusion hybrid. Next, we consider the fissile fuel breeding capacity of such a reactor. Recall the schematic depiction of the d-t fusion-fission system as suggested in Fig.15.4. The total fusion reaction rate is given by

$$R_{dt}^{*} = \int_{V_c} N_d N_t <\sigma v>_{dt} d^3 r \tag{15.14}$$

with the integration carried out over the fusion core volume V_c. Similarly, we define the neutron-nucleus interaction of type i [i: a (absorption), c (capture), t (tritium breeding)] involving nuclei j [j: g, f, ℓ, x] of density N_j according to

$$R_{nj,i}^* = \int_{V_b} \int_{v_n} \sigma_{nj,i} N_j \phi dv_n d^3r \tag{15.15}$$

where ϕ is the neutron flux depending on v_n and the integration is carried out over the entire neutron speed range and the blanket volume V_b.

A fissile fuel recycle fraction ε_r, representing the fraction of bred fuel burning in situ in the blanket, is now introduced. Then, $1-\varepsilon_r$ is the fraction of bred fuel made available for external fission reactors. For every X units of fuel bred, only the fraction $\varepsilon_r X$ is retained and subsequently fissions in the blanket. For steady state operation, it will be seen below that the fissile breeding ratio is conveniently identified to equal $1/\varepsilon_r$.

Our objective now is to find a suitable expression for the hybrid breeding ratio

$$C_{hyb} = \frac{\text{rate of fissile fuel bred}}{\text{rate of fissile nuclei destroyed}} = \frac{R_{ng,c}^*}{R_{nf,a}^*} \tag{15.16}$$

in terms of lumped reactor physics parameters. To do this we must formulate reaction rate equations which describe the production and removal of the various nuclear species of interest in the hybrid reactor. The rate equations for the total populations of tritium N_t^* in the system, and of fissile fuel N_f^* and neutrons N_n^* in the blanket are given by

$$\frac{dN_t^*}{dt} = \begin{pmatrix} \text{tritium production rate} \\ \text{by neutron capture in} \\ \text{lithium in the blanket} \end{pmatrix} - \begin{pmatrix} \text{tritium destruction rate} \\ \text{by d - t fusion in the core} \\ \text{of the fusion reactor} \end{pmatrix} \tag{15.17a}$$

$$= R_{n\ell,t}^* - R_{dt}^*$$

$$\frac{dN_f^*}{dt} = \begin{pmatrix} \text{fissile nuclei production rate} \\ \text{by neutron capture in} \\ \text{fertile nuclei in the blanket} \end{pmatrix} - \begin{pmatrix} \text{fissile nuclei destruction rate} \\ \text{by neutron absorption in} \\ \text{fissile nuclei in the blanket} \end{pmatrix}$$

$$- \begin{pmatrix} \text{rate of fissile fuel} \\ \text{removed from the blanket} \\ \text{for external processing} \end{pmatrix} \tag{15.17b}$$

$$= R_{ng,c}^* - R_{nf,a}^* - F_{-f}^*$$

$$= \varepsilon_r R_{ng,c}^* - R_{nf,a}^*$$

$$\frac{dN_n^*}{dt} = \begin{pmatrix} \text{rate of d - t fusion neutrons} \\ \text{entering the blanket} \end{pmatrix} + \begin{pmatrix} \text{net neutron multiplication} \\ \text{rate by neutron absorption} \\ \text{in multipliers of type x} \end{pmatrix}$$

$$+ \begin{pmatrix} \text{net neutron multiplication} \\ \text{rate by neutron absorption} \\ \text{in fissile nuclei in the blanket} \end{pmatrix} - \begin{pmatrix} \text{net neutron destruction rate} \\ \text{by neutron absorption in} \\ \text{fertile nuclei in the blanket} \end{pmatrix}$$

$$- \begin{pmatrix} \text{net neutron destruction rate} \\ \text{by neutron absorption} \\ \text{in lithium in the blanket} \end{pmatrix} - \begin{pmatrix} \text{rate of neutrons leaking} \\ \text{from the blanket} \end{pmatrix}$$

$$= bR_{dt}^* + \sum_x \left(\eta_x - 1 \right) R_{nx,a}^* + \left(\eta_f - 1 \right) R_{nf,a}^* - \left(1 - \eta_g \right) R_{ng,a}^* - \left(1 - \eta_\ell \right) R_{n\ell,a}^* - R_{-n}^* .$$

$$(15.17c)$$

Here b represents the blanket coverage factor introduced in Sec. 13.7 and η_j denotes the average number of neutrons emitted per neutron absorbed in j-nuclides. Obviously $\eta > 1$ for neutron-multiplying and fissile nuclei, while the fertile materials feature $\eta_g < 1$ and $\eta_\ell < 1$, respectively.

The case of steady-state operation, defined by

$$\frac{dN_t^*}{dt} = \frac{dN_f^*}{dt} = \frac{dN_n^*}{dt} = 0 , \qquad (15.18)$$

leads therefore to the following conditions on the reaction rates:

$$R_{n\ell,a}^* = R_{n\ell,t}^* = R_{dt}^* \qquad (15.19a)$$

$$R_{nf,a}^* = \varepsilon_r R_{ng,c}^* \qquad (15.19b)$$

$$R_{ng,c}^* = bR_{dt}^* + \sum_{j=x,f,\ell} \left(\eta_j - 1 \right) R_{nj,a}^* + \sum_{m=f,2n,3n} \left(v_{g,m} - 1 \right) R_{ng,m}^* - R_{-n}^* \qquad (15.19c)$$

where in Eq. (15.19c) we have substituted

$$\eta_g = \frac{\sum_m v_{g,m} R_{ng,m}^*}{R_{ng,a}^*} \qquad (15.20)$$

with $v_{g,m}$ indicating the average number of neutrons released per neutron-multiplying reaction of type m [m: f, 2n, 3n] in fertile nuclei g. The set of Eqs. (15.19a) - (15.19c) can be manipulated to give the following explicit expression for the hybrid breeding ratio, Eq. (15.16):

$$C_{hyb} = \frac{1}{\varepsilon_r} = \left(\eta_f - 1 \right) + \sum_x \left(\eta_x - 1 \right) \frac{R_{nx,a}^*}{R_{nf,a}^*} + \sum_{m=f,2n,3n} \left(v_{g,m} - 1 \right) \frac{R_{ng,m}^*}{R_{nf,a}^*} - \frac{R_{-n}^*}{R_{nf,a}^*} , (15.21)$$

where we have assumed a blanket design such that $b \approx 1 - \eta_\ell$. The first term on the right hand side of Eq. (15.21) resembles the breeding ratio of an "infinite" fast breeder, but here its numerical value may be much larger because the neutron spectrum in the hybrid blanket driven by 14.1 MeV d-t fusion neutrons can peak at a higher energy. Indeed, for neutron energies up to 14 MeV, the energy dependence of η_f is given by

$$\eta_f \approx a_1 + a_2 E \qquad (15.22)$$

where a_1 is in the range 1 to 1.5 and $0 < a_2 < 0.2$–depending upon the isotopes involved–for E in units of MeV. Employing a neutron multiplier, which obviously features $\eta_x > 1$, will provide for an increased breeding ratio. Further neutron gain takes place by fast fissions of fertile nuclei and inelastic scattering from them associated with $v_{g,f} \approx 3$, $v_{g,2n} = 2$ and $v_{g,3n} = 3$, respectively. All these neutron production mechanisms together are capable of maintaining a high neutron availability-for-breeding against parasitic absorption and leakage out of the blanket. Hence, the breeding ratio for such a fusion-fission hybrid can substantially exceed the breeding ratios obtainable with fast breeder fission reactors.

Indeed, for recent hybrid concepts designed to stress fuel production, it is estimated that a single hybrid can support up to 10 fission reactors each having a thermal power level equal to that of the hybrid; in contrast, the typical fast breeder may typically support about one companion fission reactor of comparable power. Thus, the roles of such systems would be quite different: the hybrid focuses upon fuel breeding whereas the fast-fission breeder emphasizes energy production with some breeding.

15.5 Satellite Extension

The hybrid reactor discussed above is characterized by a blanket which sustains both fissile fuel breeding as well as fission reactions. This latter function assigns some fission reactor characteristics to the blanket and consequently may impose similar safety considerations such as possible criticality or loss-of-coolant accidents and fission product release. In order to minimize these problems, it is possible to design a blanket which tailors the neutron spectrum in order to maximize fissile fuel breeding processes. The bred fuel could then be used in various companion fission reactors so that the fusion breeder reactor could be viewed as a "nuclear fuel factory" much as uranium mining and enrichment plants presently serve this function. Indeed, some additional fuel service features could be introduced if desired. For example, the fuel could be enriched in the fusion neutron-driven blanket to a desired level with a minimum of fission product accumulation–that is, rejuvenated in-situ while it is still retained in its cladding–for direct insertion into a fission reactor.

Another appealing extension can be conceived of by drawing upon the d-d rather than the d-t fusion reaction. Though the plasma conditions become more demanding, the absence of tritium fuel breeding in the blanket and handling does render this fusion cycle very appealing and makes all of the fusion neutrons potentially available for breeding. The scheme is the following.

We begin by recalling the d-d and d-h fusion reactions in addition to the base d-t reaction. If only deuterium fuel is supplied and if the reaction product tritium is consumed at its rate of production, then the overall fusion reaction cycle may be represented by

$$d + d \rightarrow t + p$$
$$d + t \rightarrow n + \alpha$$
$$\underline{d + d \rightarrow n + h}$$
$$5d \rightarrow 2n + \alpha + p + h .$$

(15.23)

In the above processes, the charged products α and p will be, to a large extent, retained in the plasma and the two neutrons will enter the blanket and breed fissile fuel. While the energetic α and p serve only for plasma heating, the bred helium-3 (h) can either be recirculated in the fuel to provide added energy release via d-h reactions or it could be extracted and used as fuel for small fusion reactors optimized for the reaction

$$d + h \rightarrow p + \alpha .$$

(15.24)

The appeal in this kind of d-h satellite fusion reaction is that the fuels and reaction products are not radioactive and, additionally, neutrons are produced only diminutively by d-d side reactions so that the chamber walls would suffer less activation. Further, the energy associated with the charged particle reaction products would be suitable for transformation into electricity by direct conversion techniques.

These features suggest that small d-h fusion reactors might be placed near populated sites to provide radiologically cleaner and smaller size nuclear energy sources. One difficulty with the satellite approach, however, is that the amount of ^{3}He bred by the hybrid is limited so that the satellite power would only be a fraction of that supplied by the hybrid-client reactor complex. Nevertheless, this could be attractive for specialized applications requiring small electrical plants.

Additionally, and as illustrated in Fig.15.6, one may conceive of one large central parent d-d reactor simultaneously breeding fissile and fusile fuel for a distributed system of various fission and fusion satellite reactors.

Problems

15.1   If fissile fuel burning were to be incorporated as a fourth layer in Fig.15.2, where would it be most effectively inserted based on neutron energy considerations?

15.2   Confirm the correctness of the $Q_{fi} / (1 - C_{fi})$ term leading to Eq.(15.5).

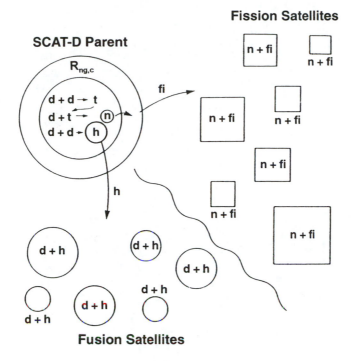

Fig. 15.6: Schematic depiction of a d-d "parent" fusion reactor supplying fissile fuel to fission satellites and helium-3 fuel to fusion satellites.

15.3   Confirm that Eq.(15.21) follows from Eqs.(15.19).

15.4   Undertake a power balance analysis for the Symbiont/Satellite system of Fig.15.6.

15.5   Estimate the d-h satellite to hybrid power ratio for the system described in Fig.15.6.

16.   Concepts and Systems

In some of the preceding chapters we have referred to specific fusion systems and devices which are currently under extensive development. Our interest here is to examine some special system concepts deserving further examination.

16.1   Direct Energy Conversion

By way of introducing some unique possibilities for fusion energy conversion, we consider the sequence of energy transformations in a fission reactor. We recall that a fission event produces energetic and massive fission products which transfer their kinetic energy by collision to the host atoms in the fuel. This heats the fuel to elevated temperatures with the resultant thermal energy then transported to the coolant which is subsequently used to produce steam under pressure. The expanding steam causes rotation of a steam turbine which is directly connected to an electrogenerator to produce electricity for a distribution network.

As previously shown, the majority of reaction products in advanced fusion cycles (d-h, d-d) are ions. The motion of these ions constitutes a current flow which could, in principle, be converted into electrical energy. Some fusion reactors, particularly magnetic mirrors, are especially well suited for such purposes because ions leaking through the mirror ends already possess desirable directional properties. Specifically then, the direct collection of charged particles represents a transformation of the kinetic energy thereof to electrostatic potential energy which can act to sustain a current through an external load, R_L. This concept is illustrated by the idealized collector shown in Fig.16.1 where an ion beam impinges on a single plate collector held at voltage V^+.

For analytical purposes we consider an ion beam with an initial angle-energy distribution such that the differential current $J(\mu_o, E_o)$ gives the number of ions with direction cosine μ_o (i.e., $\mu_o = \cos \theta$) and energy E_o crossing the plane at $x=0$ per cm^2 per second per unit direction cosine and per unit energy. Then, the total beam current J_b, in units of cm^{-2}s^{-1}, is given by integration over all ion energies and direction cosines:

$$J_b = \int_{E_o=0}^{\infty} \int_{\mu_o=1}^{0} J(\mu_o, E_o) d\mu_o dE_o \,. \tag{16.1}$$

Ions of charge q that have an improper initial direction, or too low an initial energy, are turned around by the potential V^+ prior to reaching the collector and thus become so-called retrogrades. Under the assumption that space-charge effects do not distort the ion trajectories significantly, the current J_c reaching the collector is given by

$$J_c = \int\limits_{E_o = qV^+}^{\infty} \int\limits_{\mu_o = 1}^{\mu_c(E_o)} J(\mu_o, E_o) d\mu_o dE_o \tag{16.2}$$

where $\mu_c(E_o)$ is the smallest direction cosine that an ion of initial energy E_o can have and still be collected, and V^+ is the plate voltage. For simplicity, consider a parallel beam, $J(\mu_o, E_o)$ so that Eq.(16.2) reduces to

$$J_c = \int\limits_{qV^+}^{\infty} J(E_o) dE_o \tag{16.3}$$

and the possible power generated in sustaining a load is thus

$$P_c = qV^+ \int\limits_{qV^+}^{\infty} J(E_o) dE_o. \tag{16.4}$$

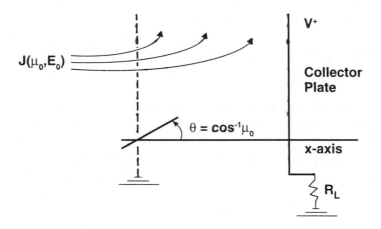

Fig. 16.1: Geometry for an idealized one-dimensional single-plate collector.

With q and $J(E_o)$ specified by the fusion device, the operating voltage V_{op} should be such that the power output is maximized:

$$\left. \frac{dP_c}{dV^+} \right|_{V^+ = V_{op}^+} = 0. \tag{16.5}$$

Performing the differentiation in Eq.(16.4) gives

$$\left\{ q \int\limits_{qV^+}^{\infty} J(E_o) dE_o + qV^+ \left[\frac{d}{dV^+} \int\limits_{qV^+}^{\infty} J(E_o) dE_o \right] \right\}_{V^+ = V_{op}^+} = 0. \tag{16.6}$$

This last term requires differentiation of the lower limit of the integral leading to

$$\frac{d}{dV^+} \int_{qV^+}^{\infty} J(E_o)dE_o = -q \, J(qV^+) . \tag{16.7}$$

Substituting this expression in Eq.(16.6) gives the following integral conditions on V_{op}^+ as the maximum power output:

$$qV_{op}^+ \, J(qV_{op}^+) = \int_{qV_{op}^+}^{\infty} J(E_o)dE_o . \tag{16.8}$$

The maximum ion beam-to-electricity conversion thus occurs at the operating voltage V_{op}^+. The efficiency, given by the ratio of power output at V_{op}^+ divided by the incident beam power, gives therefore

$$\eta_{be} = \frac{P_c(V_{op}^+)}{P_{beam}} = \frac{qV_{op}^+ \int_{qV_{op}^+}^{\infty} J(E_o)dE_o}{\int_{0}^{\infty} J(E_o)E_o dE_o} . \tag{16.9}$$

Note that $\eta_{be} \to 1$ as $J(E_o) \to J_o\delta(E_o - qV_{op}^+)$.

For a possible mirror reactor we may take

$$J(E_o) = \begin{cases} J_o \, , & E_1 \le E_0 \le E_2 \\ 0 \, , & \text{otherwise} \end{cases} \tag{16.10}$$

and assume a wide energy spread such that $E_1 \le E_2/2$. The optimum voltage is then found from Eq. (16.8) to be $E_2/(2q)$, while the maximum efficiency is

$$\eta_{be} = \frac{\frac{1}{2}}{1 - \left(\frac{E_1}{E_2}\right)^2} \, , \quad E_1 \le \frac{E_2}{2} . \tag{16.11}$$

Thus, for the relatively narrow energy spread of $E_1 \sim E_2/2$, the efficiency is ~67%, while at the other extreme, when $E_1 \sim 0$, the efficiency decreases to 50%.

While the efficiency for a single plate is attractive, values exceeding 90% are possible with multiple-plate collectors, even with a wide range of beam energies. This is possible because the voltages on various plates can be set so as to efficiently intercept particles having different energies, Fig.16.2. Efficiencies over 90% require 5 or more plates. However, a number of non-ideal effects not considered here, such as leakage currents, secondary electron emission currents and space charge effects can cause lower efficiencies in practice.

Direct collection requires extraction of a beam of charged particles from the fusion plasma. With a mirror-reactor, the escaping plasma, Fig.16.3, is first magnetically expanded in order to convert $v_\perp$ energy to $v_\parallel$, and thus form a directed flow of ions.

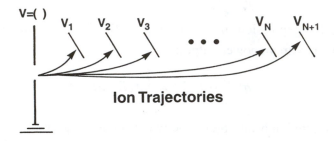

Ion Trajectories

Fig. 16.2: Voltage-plate scheme for a multiple-plate collector. The particles with energies between 0 and qV_1 do not reach a collector but those with energies between qV_N and qV_{N+1} are collected on the N'th plate.

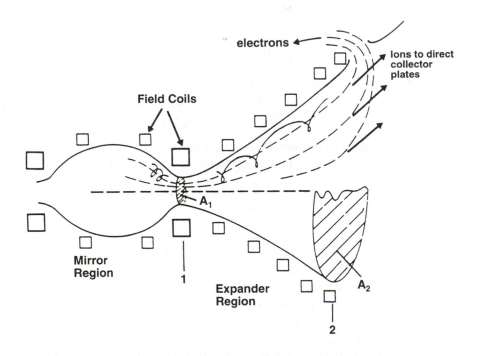

Fig. 16.3: Conversion of ion energy using a magnetic expander attached to a mirror reactor. The sharp bend in field lines exiting the expander is designed to separate the electrons from the ions.

Consider the idealized expander of Fig.16.3 characterized by an entrance with an average magnetic field B_1 and cross sectional area A_1, and exit values of

B_2 and A_2. Conservation of the magnetic flux and particle flow provides the relations

$$B_1 A_1 = B_2 A_2 \qquad (16.12a)$$

$$J_1 A_1 = J_2 A_2 \qquad (16.12b)$$

and the imposition of adiabatic invariance gives

$$\frac{E_{\perp,1}}{B_1} = \frac{E_{\perp,2}}{B_2} . \qquad (16.12c)$$

Here J_x is the particle current at position x and $E_{\perp,x}$ is the corresponding particle energy associated with perpendicular orbital motion. The expander efficiency η_e for conversion to parallel energy is evidently

$$\eta_e = \frac{E_{\|,2}}{E_o} = 1 - \frac{E_{\perp,2}}{E_o} \qquad (16.13)$$

with $E_o = E_{\perp,x} + E_{\|,x}$ accounting for the total ion energy. For a desired efficiency, η_e, the expander size ratio, A_2/A_1, is then obtained from these relations as

$$\frac{A_2}{A_1} = \frac{\alpha}{1 - \eta_e} \qquad (16.14)$$

where α is the initial fractional energy in the perpendicular direction, $E_{\perp,1}/E_o$. For the mirror example, escaping particles are nearly reflected so that $E_{\perp,1} \sim E_o$. Then, to a good approximation, $\alpha \sim 1$ and

$$\eta_e \approx 1 - \frac{A_1}{A_2} . \qquad (16.15)$$

After most of the particle energy is converted to $E_\|$ in the expander it is necessary to separate ions and electrons prior to collection. One technique is to sharply bend the magnetic field lines at the exit of the expander. Electrons will still be trapped on the lines while the ions, due to their larger momentum, will non-adiabatically cross field lines thus providing the desired separation. Separate collectors can be used for the ions and the electrons, which is particularly essential for the ions since they carry most of the energy.

16.2   Electromagnetic Coupling

The principle of magnetic induction is based on the phenomenon that a time varying magnetic field will induce a current flow in an electrical conductor. Faraday's law provides the important relation in the form

$$V = -n \frac{d\Psi_m}{dt} \qquad (16.16)$$

where n is the number of identical turns, V is the induced voltage across the conductor ends, and $\Psi_m(t)$ is the spatially integrated magnetic flux crossing the coil area A

$$\Psi_m(t) = \int_A \mathbf{B}(t) \cdot d\mathbf{A} .\qquad(16.17)$$

Here, $\mathbf{B}(t)$ is the time varying magnetic field and $d\mathbf{A}$ is the normal differential area. Evidently, a pulsed fusion reactor is necessary.

A schematic of how the magnetic field in a plasma could be coupled to an external circuit is suggested in Fig.16.4. The fundamental energy transformation mechanism is that any expansion of the confined plasma, working against the magnetic field, appears as an induced voltage across the coils, Eq.(16.16).

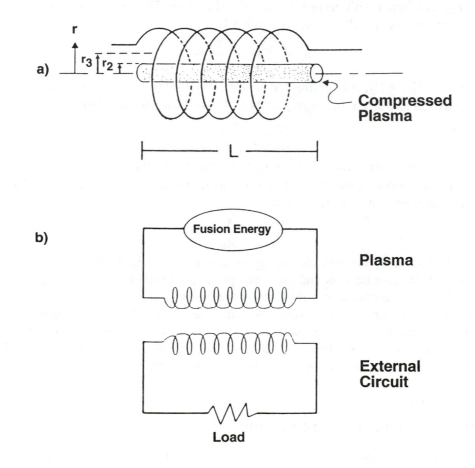

Fig. 16.4: Depiction of magnetic coupling by compression-expansion of a long cylindrical plasma. Part (a) represents the plasma-coil configuration while part (b) suggests its equivalent circuit.

Two versions of conceivable systems are envisaged. In one, the plasma is first compressed magnetically to ignition with the expansion of the plasma against the magnetic field providing for energy transformation by induction. A related approach is to move the fusion plasma through a linear burn chamber and plasma energy could be removed by having the plasmoid exit through an expanding section of magnetic field.

For either case, the analysis of electromagnetic coupling follows from the equivalent circuit of Fig.16.4. The power induced in the secondary magnet coils of n turns is given by

$$P^* = VI = -nI \frac{d\Psi_m}{dt} - I^2 R_c \qquad (16.18)$$

where $d\Psi_m/dt$ is the total rate of change in magnetic flux caused by the expanding plasma, I is the induced current, and $I^2 R_c$ represents Joule-heat losses in the coil. Neglecting Joule-losses, the energy transferred during an expansion with the plasma radius varying from r_2 to r_3 in Fig.16.4 follows by integration

$$E^* = -n \int_2^3 I \, d\Psi_m . \qquad (16.19)$$

Consider now an isobaric expansion in which a constant magnet current I is required to maintain a constant field B and in which the magnetic field associated with the coils is

$$B = 4\pi n \frac{I}{L} \qquad (16.20)$$

with L as the plasma length. The magnetic flux in the region between the plasma and coil is

$$\Psi_{mj} = \pi \left(R^2 - r_{pj}^2 \right) B , \qquad (16.21)$$

where the subscript j signifies the radius position so that the corresponding plasma volume is given by

$$V_{pj} = \pi r_{pj}^2 L . \qquad (16.22)$$

Combining these relations we find that the energy transferred to the secondary coil to be

$$E^* = -nI \left(\Psi_{m3} - \Psi_{m2} \right) = \frac{B^2}{4\pi} \left(V_{p3} - V_{p2} \right) . \qquad (16.23)$$

To understand the origin of the transferred energy E^*, we consider the various individual components of the expansion work. To simplify the analysis, we take β–the ratio of plasma to magnetic field pressures–as a constant. The expansion work q_+^p done by the plasma pressure p_p is then

$$q_+^p = \int_2^3 p_p dV = \beta \frac{B^2}{2\mu_o} \left(V_{p3} - V_{p2} \right) \qquad (16.24)$$

where we have used Eqs. (9.10) and (9.11) to substitute for the plasma pressure. However, the trapped field B_p in the plasma simultaneously expands and performs work against the confining field. This work is

$$q_+^B = \int_2^3 \frac{B^2}{2\mu_o} dV = (1-\beta)\frac{B_p^2}{2\mu_o}(V_{p3} - V_{p2}) \tag{16.25}$$

where we have introduced the relation $B_p^2 = (1-\beta)B^2$ reflecting the condition that the confining field B has to balance the total of kinetic pressure, p_p, and the trapped-magnetic-field energy density, $B_p^2/(2\mu_o)$, in the plasma. In addition, as the plasma and trapped field expand, it displaces the confining field originally contained in the volume between r_3 and r_2. The corresponding energy E_d^* must appear in the secondary coil along with pdV-work; this term is given by

$$E_d^* = \frac{B^2}{2\mu_o}(V_{p3} - V_{p2}). \tag{16.26}$$

The total energy transferred to the coil is then the sum of these three components:

$$E^* = q_+^P + q_+^B + E_d^* \tag{16.27}$$

which, upon substitution of Eqs.(16.24) through (16.26), agrees with Eq.(16.23).

Similar expressions can be derived for all the components in a cycle; expansion steps transfer energy to the coil, while compression steps require an input energy. Note, however, that the net output will be determined by the plasma-work terms Eq.(16.24) alone because the energy associated with the trapped and displaced magnetic fields must cancel out in a complete cycle. That is, in order to return to the starting point, the displaced field energy transferred to the coils during expansion steps must be replaced during compression steps. For this reason, we will compute work terms based on the plasma term alone. It must be remembered, however, that the actual energy transferred during any given stage will exceed the plasma term; this becomes important in the actual sizing of coils and the evaluation of joule losses.

As illustrated in Fig.16.5, a Carnot cycle consists of two isothermal and two adiabatic processes. The term "adiabatic" is used here in a dual sense: first, we assume no heat-transfer or radiation losses occur during the compression or expansion legs; second, since the Carnot cycle is by definition an equilibrium cycle, these stages use slow reversible processes where magnetic-field lines remain "frozen" in the plasma.

A stable plasma with an ideal beta~1 confinement is assumed throughout. Furthermore, the number of plasma particles is taken to be constant, corresponding to negligible leakage and burnup; energy carried off by neutrons and radiation is omitted, and additional energy losses such as Joule-heating losses in the magnetic coils are also neglected.

The cycle begins at point 1, where a total of $2N^*$ plasma particles ($N_i^* = N_e^* = N^*$) are assumed to be trapped in the magnetic field. The following steps then follow:

Stage 1 to 2:   Slow-adiabatic compression to a maximum field value B_{max} and temperature T_{max}. A compression work input q. is required. In the present idealized cycle, any fusion input is neglected prior to ignition at point 2.

Stage 2 to 3: <u>Fusion stage</u>. A constant temperature is maintained during the fusion burn (q_f^* = fusion-energy input to the plasma) by expanding the plasma. Thus, work q_+' is performed against the confining field.

Stage 3 to 4: <u>Slow-adiabatic expansion</u> to B_{min}, extracting work q_+. It is assumed that the temperature and density are reduced so rapidly that fusion input can be neglected after point 3.

Stage 4 to 1: <u>Cooling stage</u>. Radiative cooling q_r^* is achieved at constant temperature T_{min} by simultaneously compressing the plasma ($q_-' =$ work input).

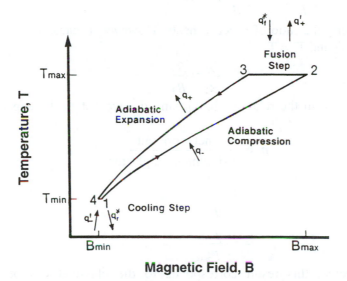

Fig. 16.5: Carnot cycle for a thermonuclear plasma. The equivalent plasma temperature is shown as a function of the external magnetic field.

During the fusion burn (stage 2 to 3), the plasma expansion work must just balance the fusion input so that isothermal conditions are maintained. To a good approximation the plasma pressure-temperature-volume relations can be taken from ideal gas laws with care taken to incorporate the trapped-field defined β parameter. This gives

$$q_f^* = q_+' = \int_2^3 P_p \, dV_p$$

$$= 2N^* k T_{max} \, \ln\left(\frac{V_{p3}}{V_{p2}}\right).$$

(16.28)

However, since pressure balance requires that $B^2 \propto V^{-1}$ during this stage, Eq.(16.28) can be written as

$$q_f^* = 4N^*kT_{max} \ln\left(\frac{B_2}{B_3}\right). \tag{16.29}$$

In a similar fashion, we find for the cooling stage (4 to 1),

$$q_r^* = 4N^*kT_{min} \ln\left(\frac{B_1}{B_4}\right). \tag{16.30}$$

During the slow-adiabatic compression (1 to 2) and expansion (3 to 4) stages, ideal gas laws apply so that

$$\left(\frac{T_2}{T_1}\right)^{\frac{1}{2(\gamma-1)}} = \frac{B_2}{B_1}, \quad \text{and} \quad \left(\frac{T_4}{T_3}\right)^{\frac{1}{2(\gamma-1)}} = \frac{B_4}{B_3}, \tag{16.31}$$

with γ denoting the ratio of specific heats. However, because of the isothermal stages, $T_3 = T_2$ and $T_4 = T_1$,

$$\frac{B_1}{B_2} = \frac{B_4}{B_3}. \tag{16.32}$$

Using this result in the relations for q_f^* and q_r^*, we obtain the expansion cycle efficiency

$$\eta_c = \frac{\text{net work out}}{\text{fusion input to plasma}}$$

$$= 1 - \frac{q_r^*}{q_f^*} \tag{16.33}$$

$$= 1 - \frac{T_{min}}{T_{max}}.$$

As expected, this result corresponds to the classical Carnot efficiency. Indeed, since T_{min} could be in the range of 10 eV while T_{max} is on the order of 10^3 eV, a Carnot limit approaching 100% is indicated. While this is encouraging, a basic question remains: How close can an actual cycle approach this ideal limit? Additional losses enter by non-ideal effects which have been neglected here. Another problem is that, for various reasons, it may be impractical to follow a Carnot cycle; for example, internal combustion engines typically utilize an Otto cycle rather than a Carnot cycle. Here, the control necessary to achieve an isothermal fusion burn may be difficult. Indeed pulsed reactor conceptual design studies to date have assumed an isobaric burn.

Another important point to stress is that the expansion cycle efficiency only applies to that portion of the fusion energy that remains in the plasma. Thus neutron and radiation energy that is processed through a thermal cycle must be included in the overall efficiency using the fractional energy flows as weighting factors with each cycle efficiency. Advanced fusion fuels with a larger fraction of energy retained in charged particles would more closely approach the electro-magnetic cycle efficiency, but even with p-^{11}B, radiation losses from the plasma would probably involve significant energy flows and be processed through a blanket thermal cycle.

A practical consideration in designing devices to operate on expansion-compression cycles is the need for a large plasma chamber in order to accommodate a relatively large expansion ratio. Thus the blanket and magnetic coils in a magnetic confinement device must have large dimensions making them relatively costly. Application to inertial confinement fusion has the advantage that the chamber dimensions are large anyway, compared to the target, in order to accommodate the micro-explosion shock effects. All of these various considerations, fuel cycle, confinement approach, economic and complexity tradeoffs, must be factored into the decision to select a best energy conversion system for a given device.

16.3   Some Emerging Concepts

Research and developmental activity towards the realization of fusion reactors has emphasized, first, magnetic confinement fusion and, second, inertial confinement fusion; low temperature fusion is a distant third. Concurrently, variations on these three as well as alternative concepts continue to appear and lead to some interest and research support. We consider two such approaches: one in the "mechanical" forcing of fusing reactions–impact fusion–and the other involving the complete transformation of the mass of the interacting particles into energy–annihilation energy.

Impact fusion is based on the notion that a macroparticle containing hydrogen atoms will–when accelerated to sufficiently high speed and made to impact upon a hydrogenous target–lead to a number of fusion reactions. An estimate of the required speed can be obtained as follows. Consider a macroparticle of mass M and speed v hitting a target. The kinetic energy of the atoms in the projectile must be of the order of the Coulomb barrier, $U_o = U(R_o)$ of Ch. 2, in order for fusion reactions to occur. For the N^* atoms in the projectile of mass M, we should therefore have

$$\tfrac{1}{2} M v^2 \approx N^* U_o , \tag{16.34a}$$

and hence

$$v \approx \sqrt{\frac{2 N^* U_o}{M}} . \tag{16.34b}$$

With m_a as the mass of each atom in the projectile, we use $M = N^* m_a$ and hence

$$v \approx \sqrt{\frac{2 U_o}{m_a}} . \tag{16.34c}$$

Then, using $U_o \sim 400$ keV and $m_a \sim 3.3 \times 10^{-27}$ kg for deuterium, yields a speed of $v \sim 6200$ km/s.

A speed of 6200 km/s is, by terrestrial standards, exceedingly high; recall that the speed of a bullet is < 1 km/s and the escape speed from the earth is ~ 11 km/s. Hence, the use of chemical explosives to attain such high speeds seems unlikely. One might, however, consider electromechanical means as suggested by the following.

Consider extending the principle of nuclear particle accelerators to macroparticles. Suppose a charge of Q is established on a projectile of mass M in a space characterized by a constant potential difference E_e per unit length. The force relation is evidently

$$M \frac{dv}{dt} = QE_e \qquad (16.35a)$$

and its integration using $ds = v \cdot dt$ leads to

$$v = \sqrt{2 \frac{Q}{M} E_e L} . \qquad (16.35b)$$

Here, L is the total path length. Currently attainable electric field strengths E_e, a reasonable length L, and assuming a constant Q/M suggest that speeds in the range of ~ 100 km/s are achievable.

A working device for such electromechanical acceleration is known as a rail gun or electromagnetic launcher, Fig.16.6. As suggested therein, a current flows through a circuit of conducting rails and movable armature to which the projectile is attached. An increasing current flow generates a time varying magnetic field which, by its Lorentz force on the conductor and for the case of rigid rails, accelerates the projectile towards a stationary target.

Numerous other concepts have been suggested to aid in high speed attainment. Among them are speed multiplication by momentum conservation, electromagnetic energy focusing and the use of losses to generate high speed ablation.

Finally, we consider "fusion" energy by annihilation.

It is interesting to note that for either fission or fusion, the mass transformed into energy represents a small fraction of that of the interacting particles; typically

$$\frac{-(\Delta m)}{m_a + m_b} < 0.01 \qquad (16.36)$$

and is thus very small.

There exists, however, one type of reaction which converts all of its mass into energy: annihilation. Annihilation occurs whenever a particle meets its antiparticle. For example, a proton p and its antiproton $\bar{p}$ combine–without any external force effects–to yield

$$p + \bar{p} \rightarrow 1456 \text{ MeV} \qquad (16.37)$$

with all subnuclear particles decaying very quickly to high energy gammas.

While the conversion of matter into energy is here complete and hence the energy release per initial mass of particles is a maximum, energy must still be supplied: antiprotons $\overline{p}$ do not "exist" naturally and have to be produced in high energy accelerators.

Thus, human ingenuity continues to be at a premium in the attainment of this ultimate source of energy.

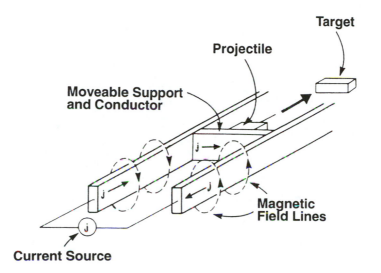

Fig. 16.6: Schematic of an electromagnetic launcher or rail gun.

PART V   APPENDICES

Appendix A   Fundamental Data and Relations

Rest Masses

Electron	$m_e = 9.1095 \times 10^{-31}$ kg	= 0.0005486 u
Proton	$m_p = 1.6726 \times 10^{-27}$ kg	= 1.007276 u
Neutron	$m_n = 1.6749 \times 10^{-27}$ kg	= 1.008665 u
Hydrogen-1 (atom)	$m_{H-1} = 1.6736 \times 10^{-27}$ kg	= 1.007825 u
Deuterium (atom)	$m_d = 3.3446 \times 10^{-27}$ kg	= 2.014102 u
Tritium (atom)	$m_t = 5.0085 \times 10^{-27}$ kg	= 3.016049 u
Helium-3 (atom)	$m_h = 5.0084 \times 10^{-27}$ kg	= 3.016029 u
Helium-4 (atom)	$m_\alpha = 6.6467 \times 10^{-27}$ kg	= 4.002603 u

Constants

Atomic mass unit	$u = 1.6606 \times 10^{-27}$ kg
Avogadro constant	$N_o = 6.022 \times 10^{23}$ mole^{-1}
Boltzmann constant	$k = 1.3807 \times 10^{-23}$ J·K^{-1} $= 8.6117 \times 10^{-5}$ eV·K^{-1}
Planck's constant	$\hbar = 1.0546 \times 10^{-34}$ J·s
Electron charge	$q_e = 1.6022 \times 10^{-19}$ C
Permeability constant	$\mu_o = 1.2566 \times 10^{-6}$ m·kg·s^{-2}·A^{-2}
Permittivity constant	$\varepsilon_o = 8.8542 \times 10^{-12}$ m^{-3}·kg^{-1}·s^4·A^2
Proton-electron mass ratio	$m_p / m_e = 1836$
Speed of light in free space	$c = 2.9979 \times 10^8$ m·s^{-1}

Conversion Relations

1 eV $= 1.6022 \times 10^{-19}$ J
   $= 4.4506 \times 10^{-26}$ kW·h
   $= 1$ kg s^{-2}·A^{-1}

1 tesla (T) $= 1$ weber·m^{-2}
   $= 10^4$ gauss (G)

Appendix B   Chart of the Light Nuclides

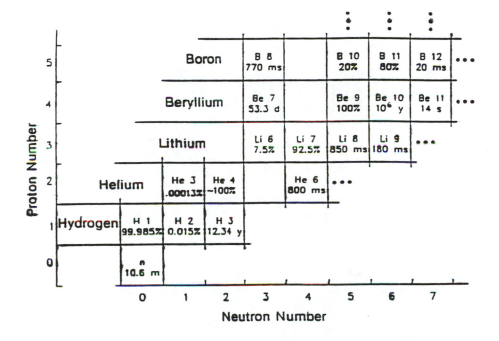

Fig. B.1: Chart of the Light Nuclides. The number identifies the natural abundance (if stable) and half-life (if radioactive).

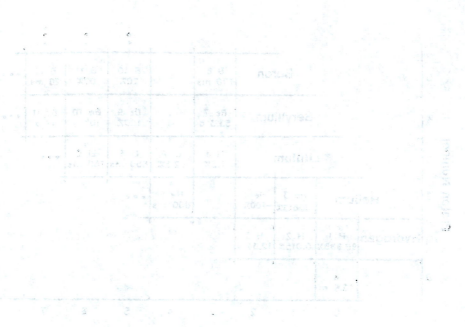

Appendix C Sigma-V and Reaction Tabulations

kT (keV)	$\langle\sigma v\rangle_{dt}$	$\langle\sigma v\rangle_{dd,t}$	$\langle\sigma v\rangle_{dd,h}$	$\langle\sigma v\rangle_{dh}$
1	6.27×10^{-27}	9.65×10^{-29}	9.66×10^{-29}	3.10×10^{-32}
2	2.83×10^{-25}	3.04×10^{-27}	3.05×10^{-27}	1.41×10^{-29}
3	1.81×10^{-24}	1.57×10^{-26}	1.57×10^{-26}	2.73×10^{-28}
4	5.86×10^{-24}	4.37×10^{-26}	4.35×10^{-26}	1.74×10^{-27}
5	1.35×10^{-23}	8.97×10^{-26}	8.90×10^{-26}	6.46×10^{-27}
6	2.53×10^{-23}	1.55×10^{-25}	1.53×10^{-25}	1.73×10^{-26}
7	4.14×10^{-23}	2.39×10^{-25}	2.35×10^{-25}	3.76×10^{-26}
8	6.17×10^{-23}	3.42×10^{-25}	3.33×10^{-25}	7.15×10^{-26}
9	8.57×10^{-23}	4.62×10^{-25}	4.48×10^{-25}	1.23×10^{-25}
10	1.13×10^{-22}	5.99×10^{-25}	5.76×10^{-25}	1.97×10^{-25}
20	4.31×10^{-22}	2.65×10^{-24}	2.41×10^{-24}	3.26×10^{-24}
30	6.65×10^{-22}	5.44×10^{-24}	4.76×10^{-24}	1.32×10^{-23}
40	7.93×10^{-22}	8.54×10^{-24}	7.28×10^{-24}	3.09×10^{-23}
50	8.54×10^{-22}	1.18×10^{-23}	9.84×10^{-24}	5.37×10^{-23}
60	8.76×10^{-22}	1.50×10^{-23}	1.24×10^{-23}	7.88×10^{-23}
70	8.76×10^{-22}	1.82×10^{-23}	1.49×10^{-23}	1.04×10^{-22}
80	8.64×10^{-22}	2.13×10^{-23}	1.73×10^{-23}	1.27×10^{-22}
90	8.46×10^{-22}	2.44×10^{-23}	1.97×10^{-23}	1.48×10^{-22}
100	8.24×10^{-22}	2.74×10^{-23}	2.21×10^{-23}	1.67×10^{-22}
200	6.16×10^{-22}	5.32×10^{-23}	4.29×10^{-23}	2.52×10^{-22}
300	4.90×10^{-22}	7.33×10^{-23}	6.00×10^{-23}	2.60×10^{-22}
400	4.13×10^{-22}	8.96×10^{-23}	7.45×10^{-23}	2.52×10^{-22}
500	3.63×10^{-22}	1.03×10^{-22}	8.70×10^{-23}	2.42×10^{-22}
600	3.28×10^{-22}	1.15×10^{-22}	9.75×10^{-23}	2.33×10^{-22}
700	3.02×10^{-22}	1.25×10^{-22}	1.06×10^{-22}	2.24×10^{-22}
800	2.83×10^{-22}	1.34×10^{-22}	1.13×10^{-22}	2.18×10^{-22}
900	2.68×10^{-22}	1.42×10^{-22}	1.18×10^{-22}	1.12×10^{-22}

Table C.1: Tabulation of sigma-v parameters (all units of $\langle\sigma v\rangle$ are in $m^3 s^{-1}$; source: J. Rand McNally, Jr., K.E. Rothe, R.D. Sharp, <u>Fusion Reactivity Graphs and Tables for Charge and Particle Reactions</u>, Oak Ridge National Laboratory, ORNL/TM-6914, Oak Ridge, TN (1979)).

$$d + t \quad \rightarrow \quad \alpha(3.5) + n(14.1)$$
$$d + d \quad \rightarrow \quad t(1.11) + p(3.02)$$
$$\rightarrow \quad h(0.82) + n(2.45)$$
$$t + t \quad \rightarrow \quad \alpha(1.26) + n(5.03) + n(5.03)$$

$t + h \rightarrow d(9.55) + \alpha(4.77)$		41 %
$\rightarrow p(5.37) + \alpha(1.34) + n(5.37)$		55 %
$\rightarrow p(10.1) + \alpha(0.40) + n(1.61)$		4 %

$$d + h \quad \rightarrow \quad p(14.7) + \alpha(3.67)$$
$$d + {}^{6}Li \quad \rightarrow \quad {}^{7}Be(0.42) + n(2.96)$$
$$\rightarrow \quad {}^{7}Li(0.763) + p(4.40)$$
$$\rightarrow \quad p(1.62) + \alpha(0.40) + t(0.54)$$
$$\rightarrow \quad \alpha(11.2) + \alpha(11.2)$$
$$\rightarrow \quad h(0.38) + \alpha(0.28) + n(1.13)$$
$$h + {}^{6}Li \quad \rightarrow \quad p(12.4) + \alpha(2.24) + \alpha(2.24)$$
$$p + {}^{6}Li \quad \rightarrow \quad h(2.30) + \alpha(1.72)$$
$$h + h \quad \rightarrow \quad p(5.72) + p(5.72) + \alpha(1.43)$$
$$p + {}^{7}Li \quad \rightarrow \quad \alpha(8.67) + \alpha(8.67)$$
$$d + {}^{7}Li \quad \rightarrow \quad n(10.1) + \alpha(2.52) + \alpha(2.52)$$
$$t + {}^{7}Li \quad \rightarrow \quad n(6.05) + n(6.05) + \alpha(1.51) + \alpha(1.51)$$
$$h + {}^{7}Li \quad \rightarrow \quad p(3.85) + \alpha(1.51) + \alpha(1.51) + n(3.85)$$
$$d + {}^{7}Be \quad \rightarrow \quad p(11.2) + \alpha(2.79) + \alpha(2.79)$$
$$t + {}^{7}Be \quad \rightarrow \quad p(4.20) + n(4.20) + \alpha(1.05) + \alpha(1.05)$$
$$h + {}^{7}Be \quad \rightarrow \quad p(4.51) + p(4.51) + \alpha(1.13) + \alpha(1.13)$$
$$p + {}^{9}Be \quad \rightarrow \quad d(0.33) + \alpha(0.16) + \alpha(0.16)$$
$$\rightarrow \quad \alpha(1.23) + {}^{6}Li(0.85)$$
$$p + {}^{11}B \quad \rightarrow \quad \alpha(2.79) + \alpha(2.89) + \alpha(2.89)$$

Table C.2: Light ion reactions. Energies of the reaction products are in units of MeV (source: J. Rand McNally, Jr., K.E. Rothe and R.D. Sharp, <u>Fusion Reactivity Graphs and Tables for Charged Particle Reactions</u>, Oak Ridge National Laboratory, ORNL/TM-6914, Oak Ridge, TN (1979)).

Bibliography

Balescu, R., *Transport Processes in Plasmas, Vol. 1 and 2*. Amsterdam: North Holland, 1988.

Barnett, C.F., et. al., *Atomic Data for Controlled Fusion*, Volume II of ORNL-5206, ORNL-5207, Oak Ridge, Tennessee: Physics Division, Oak Ridge National Laboratory, 1977.

Breunlich, W.H., et. al. (Eds.), *Muon Catalyzed Fusion*. Basel, Switzerland: Scientific Publishing, 1991.

Chen, F.F., *Introduction to Plasma Physics and Controlled Fusion, Vol. 1 and 2*. New York: Plenum Press, 1984.

Duderstadt, J.J. and G.A. Moses, *Inertial Confinement Fusion*. New York: Wiley & Sons, 1982.

Gill, R.D. (Ed.), *Plasma Physics and Nuclear Fusion Research*. London: Academic Press, 1981.

Goldston, R.J. and P.H. Rutherford, *Introduction to Plasma Physics*. Bristol and Philadelphia: IOP Publishing, 1995.

Gross, R.A., *Fusion Energy*. New York: Wiley & Sons, 1984.

Harms, A.A. and M. Heindler, *Nuclear Energy Synergetics*. New York: Plenum Press, 1982.

Hazeltine, R.D. and J.D. Meiss, *Plasma Confinement*. Menlo Park, CA, USA: Addison-Wesley, 1992.

Ichimaru, S., *Basic Principles of Plasma Physics*. Reading, MA, USA: W.A. Benjamin, 1973.

Ichimaru, S., *Plasma Physics - An Introduction to Statistical Physics of Charged Particles*. Menlo Park, CA, USA: Benjamin/Cummings, 1986.

Miyamoto, K., *Plasma Physics for Nuclear Fusion*. Cambridge, MA, USA: MIT Press, 1989.

Miley, G.H., *Fusion Energy Conversion*. La Grange Park, IL, USA: American Nuclear Society, 1976.

Motz, H., *The Physics of Laser Fusion*. New York: Academic Press, 1979.

Roth, J.R., *Introduction to Fusion Energy*. Charlottesville, VG, USA: Ibis Publishing, 1986.

Stacey, W.M., *Fusion Plasma Analysis*. New York: Wiley & Sons, 1981.

Sturrock, P.A., *Plasma Physics*. Cambridge, UK: Cambridge University Press, 1994.

Wesson, J., *Tokamaks (2nd edition)*. Oxford: Clarendon Press, 1997.

Index